PETER E. TOWNSHEND

The Text Message Marketing Handbook

Simple Steps To A Successful
Strategy for Your Business

pt3 Publishing

Brockville, Ontario, Canada

It takes 90 minutes for the average person to respond to an email. It takes 90 seconds for the average person to respond to a text message. (CTIA.org, 2011)

The Text Marketing Handbook

777 Comstock Crescent, Brockville, Ontario, Canada, K6V 6C9 613-341-8672

SAN – BUS058000 – Business & Economics/Sales&Selling/General
ISBN13 - 978-0-9866842-7-2 (Paperback)
ISBN13 - 978-0-9866842-9-6 (Kindle)
ISBN13 - 978-0-9866842-8-9 (ePub)

Library and Archives Canada Cataloguing in Publication

Townshend, Peter E., 1974-, author
     The text message marketing handbook : simple steps to a successful strategy for your business / Peter E. Townshend.

Issued in print and electronic formats.
 ISBN 978-0-9866842-7-2 (pbk.).--ISBN 978-0-9866842-8-9 (epub).-- ISBN 978-0-9866842-9-6 (Kindle)

     1. Text messages (Cell phone systems)--Handbooks, manuals, etc. 2. Telemarketing--Handbooks, manuals, etc. 3. Small business marketing--Handbooks, manuals, etc. I. Title.

HF5415.1265.T69 2013   658.8'72   C2013-905546-0
                                                                              C2013-905547-9

Table of Contents

Table of Contents 2

Introduction 4

Why Marketing With Text Messages. 6

Text Marketing 101 16

The New Marketing Model................................. 20

Success in Text Marketing... 24

How to Get Opt-Ins 31

Collecting Opt-In Subscribers. 36

Writing Successful Text Messages 43

8 Text Marketing Mistakes................................. 52

20 Successful Text Marketing Campaigns 58

Text Marketing 102 99

What You Need To Know About Short Codes 104

Text Best Practices. 112

Glossary 120

For Debbie, Luke and Maggie...
For their love and support

For Mom and Dad...
For always going above and beyond.

To Kristy...
My best editor and perfect sister.

Introduction

Over the last 20 years our society has experienced a dramatic shift in how we communicate and access information. We no longer research in encyclopedias and phone directories, and gone are the days where we carry quarters to make sure we can make a phone call in an emergency.

We have become dependent on mobile devices.

These devices (basic cell phones, smart phones, and tablet computers) have made corporations more financially stable than many countries. They have altered the way we connect, research, shop and entertain ourselves.

The adoption and growth of mobile is faster than any technology ever invented. It is because of this that we can no longer afford to ignore adapting our marketing to the mobile arena.

What this means to your business.

As a business owner, you invest a large amount of money in promoting your business. You advertise in newspapers, on the radio, and sometimes on billboards and television. You send emails, post on social media, and send direct mail.

You do this to be seen by your potential customers in hopes that they will stop, take notice of your offer, and potentially take action.

The problem is that these methods are quickly losing their effectiveness. The mass media choices (radio, newspaper, tv) are rapidly losing their audience (and your pool of potential customers) to services such as satellite radio, Netflix and news aggregators. Email is getting lost in spam filters, and social media is getting lost in the chaos.

It takes 26 hours for the average person to report a lost wallet. It takes 68 minutes for them to report a lost phone. (Unisys, 2012)

With this in mind...

How valuable would it be for your business if you could send a message to the people who are the most qualified to purchase from you, guarantee that almost all of the messages will be read, and have the highest rate of redemption?

Why Marketing With Text Messages.

Text Marketing—Using targeted text messages to a permission-based group of potential customers in order to advertise or promote a product, service, or event.

Your business doesn't need marketing... It needs more sales.

This is the bottom line truth. The only thing that's important for your business is cash flow. Without money coming in, there is no business. Bills don't get paid and your "business" is nothing more than an expensive hobby.

So, while you're business mission statement may be to help your customer solve their problems, it can never be a successful business without the necessary level of sales.

The only goal of marketing is to increase sales. Plain and simple. No matter how creative and beautiful it is, or how elegantly it "positions your brand", if it doesn't drive more business it is absolutely useless.

Traditionally, the formula for success used to be quite simple...

- You would design an ad with a headline that addressed a client's needs and desires, with a specific call to action.
- You would select a medium for the ad that would target your desired clients... Hoping that they would be able to see and remember your offer so that they would respond.
- You would address potential objections and measure response and sales rates to the specific ads to know what is working.
- You would repeat what works, and drop what didn't

Unfortunately, it is not that simple anymore. The growth of the Internet and your customer's ability to get information at their fingertips has eliminated the need to "call for more information", and entrepreneurs cannot compete with the retail box stores on price.

It is also not as easy to get the same response from the media that you advertise in. Newspapers subscriptions continue to plummet, radio is moving to satellite and digital (with some moving to a format without commercials) and local Television is pretty much a side note in the history books.

The key to increased sales and profits for your business is improving your ability to connect with your potential customers in the most effective and affordable way possible.

So how do we do this?

The Answer... Text Message Marketing

Not convinced? Here are seven reasons why your business should use Text Marketing.

1. **Everyone Carries a Mobile Phone.**

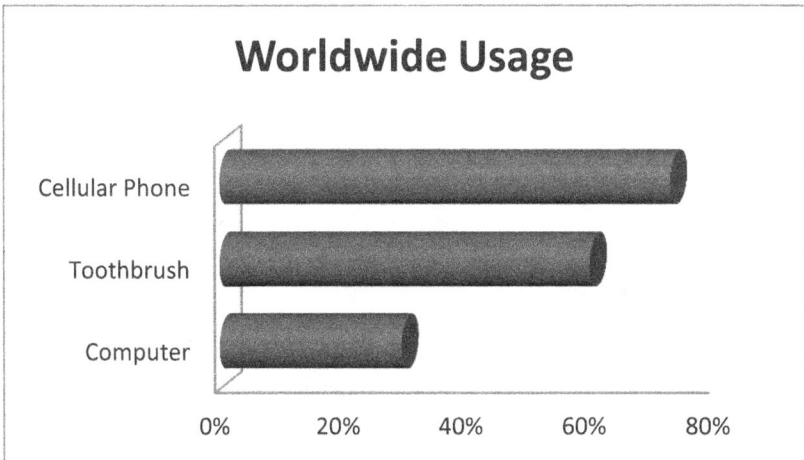

When your customers leave their home, they make sure that they have three things: Their keys, purse or wallet and their cell phones. The cell phone is really the first piece of technology that people started carrying with them all the time... and most people have these devices within arm's reach 24 hours a day. They are used to find our friends, find information and stay connected on a daily basis.

2. Text Messaging Works With All Mobile Phones

Sending text messages may sound like "old news" as far as new mobile marketing trends are concerned. After all, we have been able to send each other texts for over 20 years. Consider that while many of your customers have a smartphone, nearly every mobile phone user is capable of sending and receiving text messages. Text message marketing is the only mobile marketing approach that can connect with all of your customers, and is used more than any other medium.

Companies that can nail down their Text Message marketing strategy can open up a vast number of opportunities.

3. Text Messaging Is The Most Popular Feature of A Cell Phone

Text Messaging is everywhere and everything to consumers. According to the CTIA, 2.27 trillion text messages are sent in the US alone every year, which is 6.3 billion per day or about 20 text messages per person every day.

Top Mobile Activities

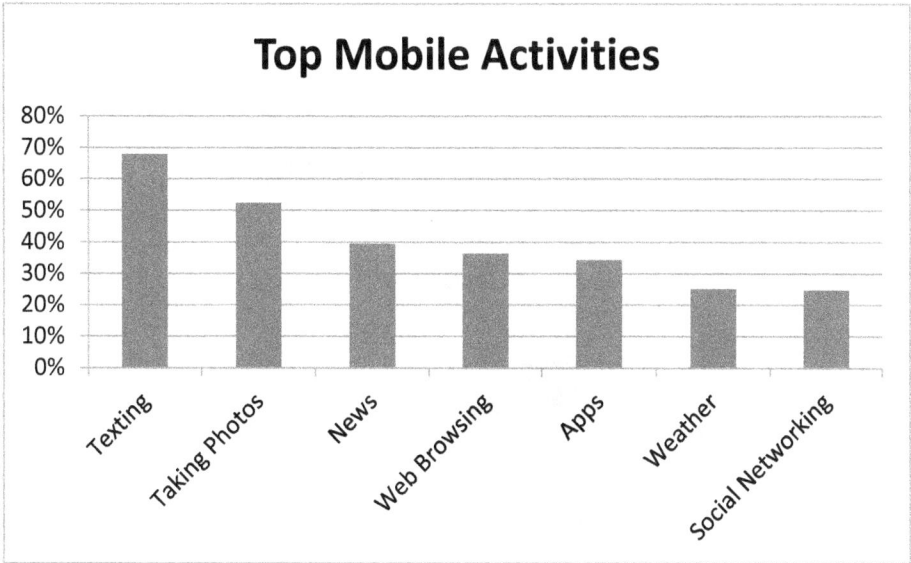

If you are worried about text disappearing as a marketing tool, you should consider that 2008 was the first year that text messages outnumbered cell phone calls. Cellular users are getting mobile text alerts seven times more than they used to. Text is not just a technology, it's one of the ways we interface with our technology and will be around for a long time.

4. Text Messages Get Read... And Responded To.

The New York Times calls Text Message Marketing "the closest thing in the information-overloaded digital marketing world to a guaranteed read."

The numbers don't lie. The open rate of text promotions/offers is a whopping 98%! Compare that to the 22% open rate of emails.

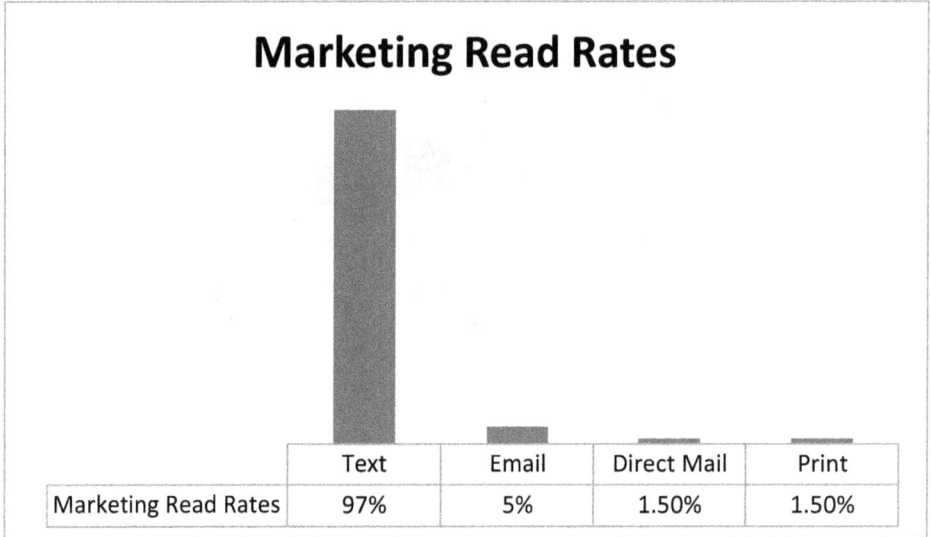

Marketing Read Rates

	Text	Email	Direct Mail	Print
Marketing Read Rates	97%	5%	1.50%	1.50%

Why?

A text message is limited to 160 characters. That means you need to say exactly what you need to say in the most efficient way possible. Whatever you're offering, you need to get to the offer quickly. Text marketing delivers your message with no filler. Just give the customer what they want, in a short and direct manner and they will respond. Text message marketing is effective and produces a redemption rate anywhere from 20% to 70%. Compare that to print, TV, radio or direct mail which averages 1.5%.

Less than 5% of marketing emails are ever opened, and even less are actually read; an unparalleled 97% of text marketing messages are read. Text messaging boasts a 14.06% CTR (click-through rate) and an 8.22% conversion rate. This dwarfs the rates of email (6.64% CTR, 1.73% conversion) and display ads (0.76%, 4.43%).

1. Text Messages Arrive...Instantly

Forget email spam filters, postage and drawing visitors into your website. When you send a text message campaign to your group of customers it arrives in seconds. Text messages are effective when communicating with customers when time is of the essence. Most text messages are read in fewer than 5 minutes after they've been received. As a result, you can count on text messaging to effectively deliver communications that create urgency.

Text Messages are reaching people at times when they aren't even expecting. It is instant and now. 90% of text messages are read within 3 minutes of reception. They connect with customers wherever they are, right when you need them to. A restaurant can send out happy hour specials directly to their customers that would be coming in after work, or florists can run the last minute Mother's Day special at 10:00 am on Mother's Day for those delinquent spouses (which I can sadly admit to being one of).

6. Text Customers Are Highly Engaged

Consumers make a careful decision on which text campaigns they choose. This differs from email campaigns, where consumers subscribe more generally to companies they may only be temporarily interested in. Because users limit themselves to businesses they truly care about when opting into a promotion, you are able to market more directly to a captive, and loyal, target audience.

Just the customers who have a positive relationship with you and want to do more business with you will receive your text messages. You only pay to market to customers who have opted-in to receive offers from you because they have given you permission to send them material.

You can maximize your response by creating highly targeted lists of potential customers (based on purchase history, demographics, interests, age groups, etc.). By targeting your campaigns to specific customers, you produce higher redemption rates and build lists that will produce immediate sales every time you send a new text advertisement to them.

This makes text messaging cost efficient. Text messages sent cost only pennies a piece, compared to direct mail pieces that can cost over $1.00.

Let's look at a quick example. Say you're the owner of a local restaurant and you've collected a list of opt-in phone numbers of your patrons. Now you want to send a message to 100 of your customers offering a free appetizer with any entrée. Say you spend $20 to accomplish all of this. Text message open rates are about 95% but let's assume only 20 of those customers actually redeem that coupon. You've just drawn in 20 new reservations for $20. If you average $50 for each reservation, your return would be $1000 for $20 in advertising.

7. Text marketing allows you to measure the results and scope of your other advertising.

Place a text call to action on your direct mail, newspaper, radio or TV ads and you will know that day which advertising is reaching your target market and which is not producing a return on your investment.

With so many customer acquisition points with text message programs, you can easily track where users are texting in to join (via in-store advertising, print advertising, etc.) and measure the relative effectiveness of that advertisement. This allows you to measure advertising in a way like never before, and is a valuable tool for marketers.

Which Ad Works Best?

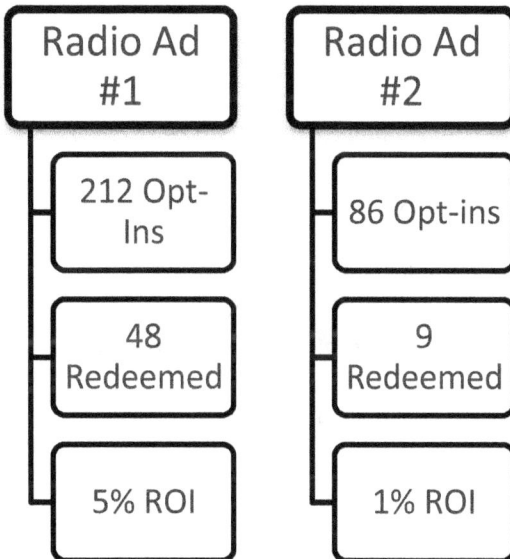

Radio Ad #1	Radio Ad #2
212 Opt-Ins	86 Opt-ins
48 Redeemed	9 Redeemed
5% ROI	1% ROI

Does it work?

Springer Mountain Farms, a family owned high-quality chicken farm, sponsors AAA Baseball in Georgia. To build their list and capitalize on their investment, they run an opt-In contest during the team's home games. At the beginning of each game, the announcer asks the attendees to send a text message keyword to a designated number in order to enter a contest to win a prize package consisting of the farm's product. A winner (or winners) is selected from these entries and the winning entries are announced before the end of the game.

The response... An average of 600 entries per game.

Imagine your list of qualified prospects after the first season?

It's not difficult to make the case for text messaging.

- Direct mail (flyers or targeted letters) - slow and often thrown in the trash without a glance
- Telephone calls - often untimely, annoying, and intrusive
- Emails - usually impersonal and disregarded as spam.
- Text messaging - popular, personal, immediate, involving, and never ignored.

Text Marketing 101

How Text Marketing Works.

Statistics show that the number one reason that business owners haven't implemented text messaging into their marketing mix is a lack of understanding of the process and how to make it fit into their current strategies.

Many believe that text message marketing is complicated, that they may be violating rules and regulations (opening them up to legal cases), or confuse this legitimate method with mobile spam.

The truth is that text message marketing is highly regulated in order to protect consumers and marketers alike. These regulations are carefully monitored by many governing bodies, mobile carriers and technology suppliers in order to prevent spamming. This ensures high response rates as you don't have to compete with hundreds of unwanted messages.

With this being said, before you can understand what text message marketing is and how it works, you must first understand what it is not. Text message marketing is not spam. You do not buy lists, or rent from someone else. It is not a one-time trial run, a magic bullet, or a quick fix. In order to make text message marketing work effectively, you must integrate it with your current marketing practice.

There are two stages of success in a text marketing program. First, you must build a permission based list of potential customers. The second stage is to monetize your list by sending ongoing promotions to your lists.

Stage 1 – Building Lists.

This first step is the most difficult, least rewarding, and the stage in which most new marketers abandon their efforts.

The reason that we market with newspaper, radio, television, and Google is that we borrow from their audience in order to try to attract their attention and get the audience to take action in our favor. Now that the media is losing their audience, the group of people we market to is shrinking. Instead, we need to become the media. We need to build our own targeted audience of customers that are interested in exactly what we have to say.

Joining a text messaging program is easy. There's no form to fill out, no website to visit. All that's needed is to use their cell phone, and send a quick text message.

A mobile phone user sees something that directs them to text a keyword to a 4-6 digit short code. For this example let's use a bakery. The instructions could say "Text DONUTS to 555444 for your chance to win free donuts for a year!" These instructions could be placed on the top of the donut boxes, read aloud in radio ads, included in newspaper ads and flyers, anywhere the bakery can place it.

The potential customer composes and sends a new text message on their phone. The "to" address will be 555444 and the message body will be "donuts."

That message is received by text message marketing platform that immediately sends back a response, like "Thank you for entering. We will draw shortly, but in the meantime enjoy a free one today. Text STOP to end."

Now the bakery knows that they have a potential customer that likes donuts. They should also build different lists for each different customer segment that they have. For example, lists of bread customers, deli customers, etc.

Building segmented lists is the key to maximizing effectiveness and response. A consumer may opt-out of any mobile-marketing campaign simply by replying "STOP" to any text message received. While opt-out rates are typically very small, you want to send messages that are in line with what the customers are interested in.

Stage 2 – Send Ongoing Promotions.

Once someone opts into your program, their cell phone becomes your billboard. Your business has just earned the most intimate advertising space; a direct connection to your market, and the ability to entice new and repeat customers with your promotions.

Now the bakery has built up its database of opt-ins, they can then begin to send out their promotions, typically once or twice per week. This should be scheduled in advance and advertised at the point of enrollment.

United States of America's President Barack Obama experienced tremendous success utilizing mobile text message marketing when he was blasting out campaign updates to nearly 3 million subscribers. Coca Cola has over a million subscribers that they engage with the My Coke rewards program. 47% of those utilizing the My Coke rewards program are 35 and older. Coca Cola is not alone as Ashley Furniture, Jiffy Lube, Papa Johns, Subway, Best Buy, Village Inn, Arby's and many other brands are successfully implementing text message marketing campaigns.

Customers that are typically engaged in traditional marketing channels are welcoming a new experiences and opportunities to engage with brands, products, and services.

The New Marketing Model...

How To Build Text Into Your Marketing

The power of text message marketing is that it is not meant to replace your current efforts, but to take the customers that you earn from your current efforts and develop the relationship to create larger purchases, repeat business and referral sales.

It is part of the new marketing model. In the new model, businesses realize that they can no longer depend on traditional marketing channels for a consistent source of new customers, as the audience becomes more dispersed across multiple channels.

In my experience as a sales manager, I realized that I had two different types of salespeople. I had salespeople who had a great knack for prospecting. These were the outgoing personalities that made the door swing and generated a large number of leads for the business. These skills were definitely an asset, but they often left money on the table as their closing rates were often lower as well as their average dollar value per sale.

Then there were the closers. These salespeople were highly skilled relationship builders and often generated more business with fewer opportunities. They were often the top earners as they maximized their dollars per sale, generated more referrals, and often converted other people's missed opportunities into new business.

Marketing works the same. Businesses often believe that marketing often ends with the customer walking through the door. Once this happens, the sales process takes over and ends with a purchase.

The new model, like the closers, insists that marketing continues after the initial contact and never ends. It means that the goal of marketing is to create initial, repeat and referral business. Traditional media has a current audience and is used and maximized to make the door swing. Text marketing integrates with traditional media in order to track response, generate contact with the consumers that might be on the fence or need more information, or deliver an incentive.

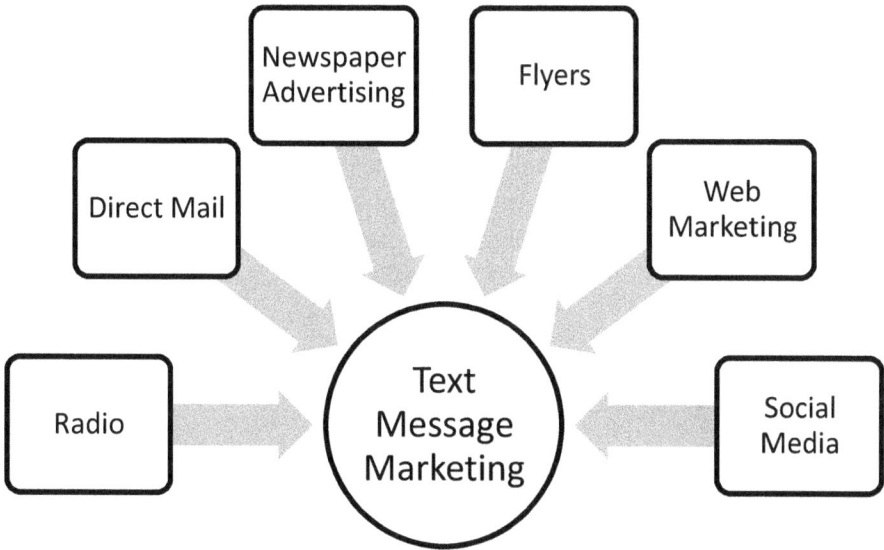

The process after initial contact used to be the domain of the sales people. This was often a lot of hard work and produced limited results. Now we use tools of social media, email marketing and text message marketing. Social media and email marketing should be used to deliver more detailed messages more frequently. You are more likely to have your message lost in the crowd or possibly even deleted as spam (albeit sometimes accidentally). Text marketing needs to be more carefully planned and used less frequently... perhaps once per week. This will ensure that it is taken more seriously and acted upon more frequently. If you want to send out text messages more often, consider creating multiple contact lists.

While I may refer to this as the new marketing model, the truth is that this model has been a secret held by the big retailers and direct mail companies for generations. They use the initial offer to generate contacts. These companies often lose money on their initial offer as the profits come from building a relationship with the customers. The customers that have trust are more likely to spend more and refer more business to you.

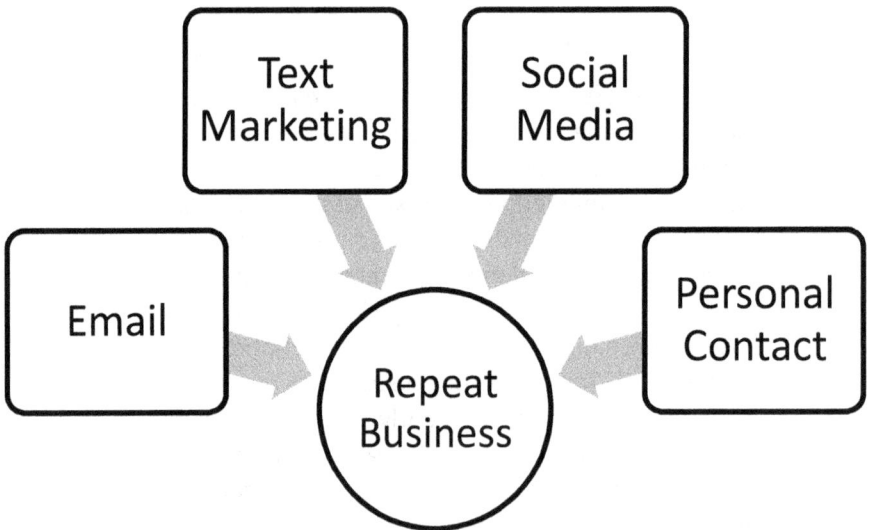

Text Marketing

Social Media

Email

Personal Contact

Repeat Business

Success in Text Marketing...

Strategy, Measurement and Resources

A clear strategy is the foundation for your text marketing efforts. When you are limited to 160 characters for your message, you need to be very clear with your expectations; what you expect to deliver to your customers, what results you expect text to deliver to you, and what you require in order to accomplish these objectives.

Studies show that business owners face challenges and barriers when approaching text message marketing. These challenges include:

- Lack of effective mobile strategy (55%)
- Inadequate staffing resources and expertise (55%)
- Budgetary limitations (45%)

It is no surprise that lack of strategy and resources are the first two hurdles that business face in incorporating text marketing. Even with these challenges, most said that they had spent more on text marketing this year than in previous years. A trend that we can be sure will continue to increase.

Defining Your Marketing Strategy

To begin crafting your mobile strategy, clearly understand the following questions:

- **What can make your customers passionate about your business**? Is your service designed to give your customers extra family time? Does your product solve an urgent problem? Can You tap into a customer's passionate hobby? Can you make your customer's more efficient and effective at their job? McDonalds provides cost-friendly meals to time-strapped families. Jiffy Lube is an oil change business designed to give customers an oil change without interrupting their busy lives. What does your business deliver?
- **Who are you looking to target with your text marketing campaign**? Do they live close to your business? What age range do they fall under? Would a specific gender be more likely to buy from you? What is his/her household income? What do they currently buy and how much do they normally pay for it?

- **Why should your customer do business with you?** What sets your business, product, or solutions apart? Pricing? Quality? Customer service? Ease of Use? Are you providing service that no one else is providing? What is your target client customer looking for when they're deciding among similar businesses? If you cannot have a different product, how can you deliver it in a unique and special way?
- **What profits are you looking to gain?** Will you be pricing your product below your competitors or above last year's model? What are you looking to gain from your products and services?
- **What specific solution will your text marketing provide for your customer?** What will you offer with your text marketing service that you're not offering with any of your other marketing channels? What will text marketing do for your business... Will it:
  - Provide better service to existing customers?
  - Attract more new buyers?
  - Gain a competitive advantage?
- **How will you let your target customers know about your text marketing program?** Will your keywords and short codes be written on business cards, fliers, websites, newsletters, or other marketing channels? Is it easy to integrate text marketing into the traditional marketing channels you're already using?

Establishing Objectives

In professional sports, teams and players rely heavily on statistics to understand their strengths and weaknesses. Building on these areas is a key factor to their success or failure. For example, in baseball the batting, pitching, and fielding statistics demonstrate a player's strengths and weaknesses. Players will then watch video, practice and develop additional skills to improve in both their strong and weak areas.

In the instance of text marketing, you need to study key statistics to enable you to optimize your text marketing program. Each consumer response provides specific and useful feedback that can be leveraged to refine your program.

Traditionally, you may measure your success entirely on return on investment. Such as, I spent $2,000 on marketing and generated $20,000 in business. While this is the ultimate aim of marketing, it does not give you the full picture. Here are some additional factors that you will want to track:

- **Opt-in rate:** In your traditional media, you may track number of appointments booked, number of visitors, or number of leads generated. You can now add number of opt-ins generated. These new subscribers represent the possible future sales for your business.
- **Opt-out rate:** When running your campaigns, there will always be a number of subscribers that decide not to continue receiving text messages

from you. This is one of the factors of doing business. A higher opt-out rate can mean that your messages need to deliver more value, be sent less frequently, or take a less aggressive sales approach.

- **Click-Through Rate:** If your text message contains a link to a mobile landing page, you should be tracking how many customers select the link and are delivered to the landing page. With this statistic you can see how many people think that the message delivered enough value in order to visit the website, and from this point, how many purchased your offer. This way you can find the weak link in your marketing chain and improve what needs to be improved.
- **Redemption/Conversion Rate:** The reason that marketing on the internet vastly improved the way that businesses advertised was that businesses could track how many people viewed their ads. This was impossible with TV/radio/newspaper. The best you could do with the traditional media is estimate based on their distribution. With text marketing, you know how many messages are distributed and you should include the ability to see how many of those messages turn into new business.

Once you have these key performance statistics in place, you will be able to establish specific goals to know whether you've been successful or not. You can determine a specific number of opt-ins, redemptions and profits that were generated from each channel. As well as which messages generated the most traffic. This information is invaluable for determining whether or not your text marketing ideas are working so you can make changes and adjust.

When planning your campaign, you also need to plan a Seventh Inning Stretch... a specific time to check your progress and make adjustments and refinements as needed. The metrics that you set should be carefully monitored to gauge if the results are what you expected. Watch for shifts in consumer behavior.

You should also establish overall business goals. Goals such as, by what percentage would your sales figures increase by building a text message marketing list? Do you expect a 10% increase in overall sales by having current customers make one additional purchase per year? Did you hope to bring in about 75 new customers just from people becoming more aware of your products and services via your text messages?

Resources Required

When talking about resources, most business owners only consider that they need to establish a financial budget for these efforts. The reality is that there are 3 costs to any marketing challenge; financial, staffing and time.

For example, social media is available for no financial cost to a business but requires significant staff resources (technical knowledge and those that know what to write in order to produce a result) and time. You need to calculate the challenges and barriers in regards to people and time in order to see whether or not the method is valid. Would you consider social media worthwhile when considering the people and time required in making it work?

Does your organization have dedicated manpower and resources for text marketing? If the answer is no, you have two choices in solving this, you can either develop and hire internal resources, or outsource and hire a company to manage this for you. It is always valuable to train and develop these skills in your internal staff, but the easiest way to get started is to find a text provider that will help with management of your program to get you started.

When calculating your resources, consider the following:

- Are you committed to continued support and growth for text marketing efforts? Have you communicated this commitment to your staff in order to ensure that they are fully prepared to help create successful campaigns?
- Where do you place responsibility for the success of text marketing in your organization?
- Can you put in place a strategy to develop the necessary staff expertise and that adequate staffing resources are available?
- Are you providing your staff the appropriate amount of time to invest in building and developing your program?

Just as a baseball team is made of players who each possess a specific set of strengths, your program's success requires skills from a range of team members.

How to Get Opt-Ins

People's mobile phone number is one of the most guarded numbers they have. Your first goal is to move a customer's concerns from, "You're going to hit me up with marketing" to "You're going to give me something exclusive and valuable."

Text message marketing requires you to ensure that your subscribers give you permission to market to them, or opt-in. You must respect the right of your customer to control which messages they receive. If a subscriber opts-in to only one program, you cannot send messages for another program or messages that don't relate to the program they opted-in for.

Your key challenge is going to be building an opt-in list from scratch. While the best way to start is to integrate the mobile messaging campaign into all other initiatives, you also need to deliver something of value in order to entice them to take part.

This is the investment involved in building an opt-in list. The payoff is delivered once the customer is opted-in and over time messaging allows you to generate a direct response from the list. This is especially true with text message marketing, where response rates can be as high as 38 percent.

What Do You Give Them.

Even though you're customers are becoming quite familiar with text messaging, it is really important to start your text relationship by giving them something. A call to action without the promise of an immediate reward will come across like spam.

There are various incentives that you can give to customers that they may find useful. The key here is to offer something that can be easily redeemed from their cell phone. You can consider offering the following things for free:

- **eBook or Series of Information:** Customers are often looking for easy to access information on how to solve their problems, and you have a topic of expertise. This will establish you as a trusted resource to your customer and give your customer a solution without having to spend hours surfing the internet. You can give a free download link, or series of text messages after someone has joined your campaign.
- **Video:** Just like an eBook or text series, you can offer a video link on a topic that will appeal to your prospective clients. The video may be related to expert advice, a how to video or something else. This has become much more popular as Twitter has even

created new apps to deliver mobile videos. Think of your customer viewing a video in your store as they are deciding on ordering a product or making a purchase.

- **Free stuff:** People love something for free, and providing a free incentive is just rewards for someone who you want to be an active participant in your mobile marketing campaigns. The benefit of providing a gateway step in texting-in to receive a free gift is that not only will clients have to come in to your business in order to claim their gift, but are also required to leave you their contact information and an open channel for future contact.

- **Contest Registration:** The ability to win a prize or participate in a contest is always a great way to build a subscriber list. It creates urgency to act quickly and allows you to control the cost of the prizes. The challenge with a mobile contest is that you need to ensure that you follow the rules and guidelines in your area. Always include a link to the rules and regulations and be very clear about prize dates and eligibility requirements.

- **Mobile Coupon:** Ten billion mobile coupons are expected to be redeemed this year, a sure sign that both retailers and consumers alike are embracing the channel. As a small business owner, mobile coupons give you an easy and affordable way to drive opt-ins and market your products and services. These coupons can be as simple as having customers show a text on their phone, or link to a mobile web page that gives a coupon that is similar to the conventional print coupons we see in magazines and newspaper inserts. Cell phones are perfect for coupons because your customers will have them in their pockets at all times. Most importantly, a mobile coupon has a redemption rate of between 10 and 44 percent. Compare that to printed coupons that have a response rate of two percent.

Here are a few tips to help you effectively generate mobile opt-ins:

1. Integrate Your Incentive with A Text Autoresponder - If you want to use incentives to attract new opt-ins, use a text autoresponder this will ensure that your customer receives a message immediately after they submit their contact information.

2. Make Your Offers Drive Action – While text messages will build relationships, remember that their primary purpose – to drive people into your business to spend money with you. People use social media to build relationships and socialize. Text is for immediate solutions to their problems and important incentives. Keep it to the point and make sure it is action driven.

3. Make Your Offers Time-Sensitive – If sending out coupon offers, contests, and freebies using text messaging, most of your audience will read the message within minutes. You can increase the responsiveness to your offers by making them time-sensitive. If your offer is worth it, your customers will respond to a 24-hour only special or to an offer available during happy hour the same day the message is sent.

4. Keep It Super Simple – In order to prevent any roadblocks for your customers, make sure your incentives, opt-in mechanisms and promotional materials are easy to read, easy to understand and easy to redeem. Also keep in mind that some of your customers will be using a regular feature phone in order to connect with you and not a smart phone. With this in mind, simplicity always wins. You don't want to cause any problems for your text subscribers when it comes to taking you up on your offers.

5. Remain Professional – It may be tempting to add a little personality or humor to your text message offers, but don't do it. Save getting more personal with your audience to your social media profiles, not on their mobile devices. Some businesses make the mistake of getting "too relaxed" with their messages, which turn-off some of your subscribers. You cannot communicate as clearly in 160 characters as you can in other media. You may risk losing business this way... don't risk it.

Collecting Opt-In Subscribers.

Now that you have your incentive in place, you have to follow some very clear policies and practices to collect opt-ins. If this process is done incorrectly, you may end up in hot water. Make sure you check with your local authority to ensure that you are following the correct processes.

Opt-In Mechanisms

There are three acceptable ways for a consumer to join you text marketing list. This list must be built; you cannot rent someone else's list or buy a generic list of numbers. These three are:

1.   The customer uses their mobile phone to join your text message database by texting a keyword to a short code. This message is considered a "M.O.", or mobile-originated message. Your text message provider records this type of opt-ins for you and maintains a file as proof of consent.

2.   The customer can join a subscriber list by filling out an online form. This form must

explicitly explain what the person will receive when providing his/her mobile number. Your provider may provide a form and automatically record this type of opt-in; you may be responsible for recording and documenting this type of subscription. Make sure that either way you understand this in case you receive a complaint or are questioned by an authority.

3. Subscribers can also join your marketing database by filling out a paper form. This form must clearly explain what the person will receive when providing his/her mobile number, as well as include appropriate legal copy. You are responsible for maintaining these records.

Note: You cannot require a consumer's consent as a condition of purchasing anything.

Required Notices

Confirming the Subscribers' Opt-in and Initial Messaging

There are very specific legal requirement for permission to market using text messages. In the early stages of text marketing, unscrupulous people would buy lists, or spam random numbers. This created numerous complaints to the regulating authorities as well as mobile carriers. It is because of this that each text messaging platform provider must have processes in place to make sure that the person signing up for mobile alerts is actually the owner of the mobile number. This makes sure that no one is entered into a program without their permission. In the case of website opt-in and paper forms, you should use a "double opt-in" process.

You can satisfy the double opt-in requirement by sending an initial text message to the mobile number asking the subscriber to reply Y or Yes to the received message. This verifies that they are in possession of the mobile phone and give permission to join the mobile database and receive text alerts.

> XYZ Burgers VIP Club,
> 5 msgs/mo. Reply Help for
> help. Msg & Data rates may
> apply. Reply Y to confirm
> your sign up or STOP 2 end
> Show this text for a free
> drink.

Tip: If you are sending a one-time message, a double opt-in is not required. If you are sending recurring messages (and why wouldn't you want to send more messages after you have earned the contact), it is highly recommended that you follow the double opt-in requirement.

Other policies that you should follow:

- If the subscribers are entering a contest or texting in for a free incentive, you need to let them know that they will be added to a marketing database and will receive additional messages. Adding some copy that says "By entering this contest you will signed up for our mobile club and receive no more than two messages per month" near the instructions is usually sufficient.
- It is always best practices to let your subscribers know how many text messages that they will be receiving and the frequency.
- You should provide a link to you privacy policies and terms of use.
- It is a requirement to provide a system for assistance. Many carriers have a keyword "Help" reserved for the ability to get assistance. It is also a good policy to provide a phone number for your customers to contact in order to get personal assistance.
- You must provide a clear procedure for subscribers to be able to revoke their permission. This is called an opt-out procedure and is typically adding the phrase "Text STOP to end" at the end of the message. Most

service providers have this procedure automated.

Where To Place Your Calls To Action

In order to capture people's cell numbers, you need some way to get their attention. There are an unlimited number of places for you to place a keyword/short code, or even a QR code to generate a mobile oriented opt-in. You want to give your clients the opportunity to join your text marketing campaign at every opportunity. Here is a list of ideas for you to start with…

- If you're a restaurant, you have tabletops. If you have a store, place notices at point of purchase and at strategic locations within your location.
- Place an opt-in form or keyword/short code on your web site
- If you are running billboards or radio spots include text calls to action.
- One company included the initial invitation to text in as part of its ads in digital jukeboxes.
- Another campaign advertised text coupons inside an EA Sports video game.
- Do you have on-site signs, end-of-aisle displays
- Using Mobile search ads. The prospective subscriber can be directed to a mobile webpage with an opt-in form.
- Add Text to Join on flyers and box toppers

- Combine with your other e-newsletter or direct mail
- Post on website and social media outlets (Facebook, Twitter, YouTube)
- Pass out business cards with your keyword and short code
- Add text tag to receipts
- Create a staff contest
- Go viral by asking your customers and subscribers to refer a friend.

The list really can be endless. The best idea that I ever encountered on this was when a company brainstormed with their staff and came up with a list of 50 ways to build their subscriber database. This list became a call to action for the staff so that when they had any free time they would refer to the list and find something that they could do at that moment to build more subscribers.

A Note On QR Codes

QR or Quick Response codes are the two dimensional specialized "bar codes" that can be scanned by smart phones to access any type of information. You can generate these QR codes to do many things, even to generate text messages. This can be handy in creating a short form way to text in the appropriate information to join your club, but you must practice caution. Here are some tips if you consider using QR Codes.

- QR Codes cannot be used from all devices – In order to access a QR Code, a client must have a smart phone and must have a specialized application on their phone to scan the code. Many subscribers will not have a smart phone, even more won't have the app or desire to scan it. QR codes are handy, but don't consider it a primary tactic in text marketing or mobile marketing in general.
- Add Instructions By Your QR Code – Customers may be intrigued by what's behind your code, but cannot scan it for some reason. Also provide your keyword and short code by the QR code and additional instructions or mobile website so anyone can participate in your offers.
- Also on the collateral, let your subscriber know the purpose of your barcode so that you can give the consumer a reason to actually scan it.
- Make sure the QR code printout has a high contrast, ideally in black and white.
- Pay attention to where the code is placed, so that it will be displayed in a place where the user will have enough time to see it and scan it. Don't place it on a vehicle, moving billboard or place that is difficult to scan.
- Test and re-test the code before printing.
- Don't put a QR code underground, in subway stations and other deep-interior structures where customers can't get the reception they need or in any location where the user has no wireless access.

Writing Successful Text Messages

You have your strategy in place, you know how you want to connect with your customers and have your technology in place. Now how do you ensure that you generate the response and influence you want in only 160 characters (or less). With so much on the line, you better craft them properly.

The Basic Formula

The good news is that there is a basic formula. This is a guideline to follow, not like the law of physics. While there are many great campaigns that don't follow this and still work, in the beginning it is best to learn the basics. When you are more comfortable and have done testing as to what your customers respond to effectively, than you can feel free to mix the first three around, but make sure you always end with a call to action.

| Identify Sender | What Are You Giving Them | Why They Want It | Call To Action |

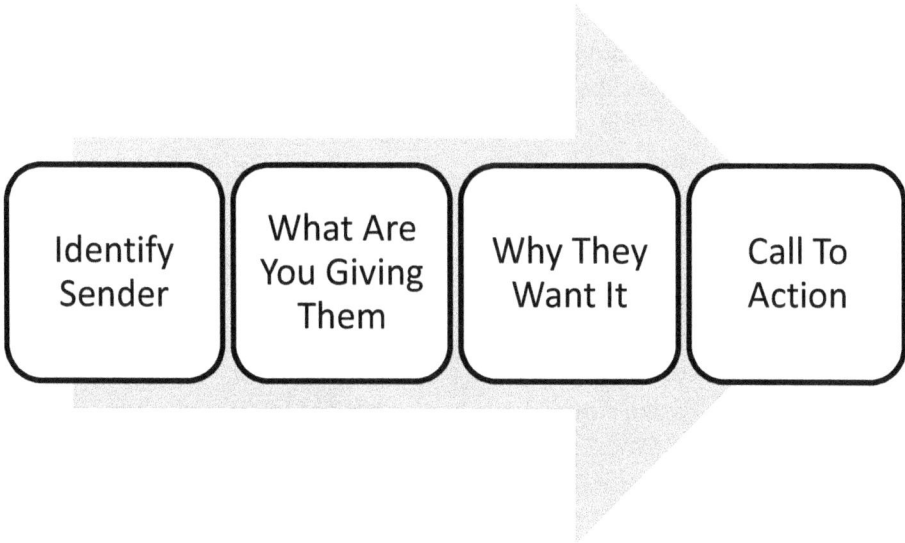

- **Identify The sender.** While your service provider might have the ability to display your name in the sender field, don't always rely on this. Many times your customer only sees your short code when a text arrives. This may seem quite simple, but make sure you identify who you are in the content of your text.
- **What You Are Giving Them**. Text Messaging is an instantaneous medium and attention spans are short. Make sure you capture attention by quickly stating your offer and making it easy to understand. Include real-time offers, gifts, discounts, coupons, mobile alerts, links to free applications, contest entries, video, and/or mobile alerts.
- **Why They Want It**. This is the "what's-in-it-for-me" factor. Customers need to know how they will

benefit immediately by taking action on your text. You should also make your customer feel that, as a subscriber to your text message campaign, they are important and getting something that others do not.

- **Call To Action.** This is the essential "what do you want the subscriber to do immediately after they read your text". While the offer is clearly important, the right the call-to-action can make or break your campaign. If you want to quote your phone number... tell the customer to dial it, and make sure there is a link for them to dial right from the message. Create a sense of urgency by adding an expiration date, and including the phrase "when you show this text message" to get people to participate while they are in your location.

Also remember that it is essential to end every message with a way to opt-out of receiving messages from your company. Typically the phrase "Text STOP to end" will work best.

Additional Tips For Writing

1. **Encourage Sharing** - To get the most from your marketing message, suggest to subscribers that they might want to forward the message to their friends.

2. **Vary Your Message** - Prevent looking like spam to your subscribers by avoiding the use of similar messages, or a cluster of messages that have a similar offer (unless it is heavily featured in the campaign. Try to make your messages vary greatly from one another and provide unique information, deals or codes.

3. **Write Short Sentences** – Text marketing by nature doesn't support long-winded descriptions of your promotions, products and services. If you tend to write longer sentences, try breaking them apart. Short sentences force you to get to the point quickly and improve the efficiency in how you write.

4. **Keep It Short And Simple** - Take one marketing message at a time and don't try to cram in too much information. If you want to make more points, write more messages.

5. **Find Great Reasons for Texting -** Try to tie your messages into something new and urgent you want to share with them: A new product launch, an event, an urgent action, a sales success, a limited time discount, social causes, seasonal messages and holidays are all great reasons to send text message.

6. **Read the headlines** – The tabloids and newspapers have done a lot of research and practice getting the right way to attract an audience's attention in short sentences. Research headlines and see what you can learn for your marketing.

7. **Use links to mobile web pages** – If your message is so important that you can't abbreviate it, than include a link in your text message that directs the subscriber to a mobile-optimized web page. If your subscriber has a feature phone without internet access, they can always type it into their computer's web browser.

8. **Watch out for technical considerations -** Occasionally, while writing text messages, you may find restrictions on the use of special characters.

Text Message Samples

Do you need inspiration to get started writing messages in 160 characters? Read through our text message sample file. These may spark ideas that you could apply to your text marketing campaigns.

Tuesday Special at Joey's Pizza: Feed the Family. 1 Large 3 Topping Pizza for $9.99. Show this text to order. Text STOP to end.

Ladies Night Wednesday! Be our VIP guest at Wilcox and get 2 for 1 cocktail specials until midnight! Show this message for no cover charge! Text Stop to end.

Club Fabulous wants you to jump the queue on Saturday night. No line and 50% off the registration for you and a friend by showing this text. Reply STOP to end.

Real Estate graduate courses. Free advanced course introductory seminar. See the intro video on your mobile phone bit.ly/regraduate Reply STOP to end

LADY FASHION members prepare for spring and receive a 50% discount on all styles tomorrow at 100 Chapel Street open 9am and show this text. Reply STOP to end

Our Sunday Treat! Enjoy a complimentary manicure with every facial booked today. Call Your Beauty Parlour at 613-555-0011 to book. Text Stop to end.

Back to school 50% off all leather laptop bags at www.abcbags.com for the next 48 hours only. Use coupon code HALFOFFTEXT at checkout. Text STOP to end.

Rubber Ducky Wine Show Special, Get 10% off any bottle of red when you stop by the Blue Duck booth. Show SMS to redeem. Expires 1/4/13. Text STOP 2 end.

Martial Arts World VIP Member Special: Get a $10 voucher when you buy more than $50 worth of gear. Show SMS to redeem. Expires 1/4/13. Text STOP 2 end.

Best BluRay deals this week on BluRaystore.com! Get 3 of the latest hit movie releases and take 1 free Live Concert disc of your favorite band. Text STOP 2 end.

Special VIP discounts on soul albums released within the last 3 years! Get your grooves at 25% off at Soul Collection when you show this text! Text STOP 2 end.

This week's VIP special offers at Fresh Foods: Mo-We: roasted chickens $3.50; Tues: soymilk $1.39; Thu-Sat: Fresh Norwegian salmon $5.39. Text STOP 2 end.

Top quality Mediterranean olives now here! Show this and get the promotional prices before we officially launch on Saturday! Greetings, Laine's Text STOP 2 end.

Happy Holiday from Mama's Deli! As a special customer you get 15% off bakery items today with this message. Dessert like Mama made. Text STOP 2 end.

Cool Promotion at BioDome! 50% off our new fat-free bio yogurt with this text and get our bio-life book with tips on how to vitalize your day. Text STOP 2 end.

Highly-qualified dietitian at Vita-store on Monday from 11am to 3pm. Free weight loss counseling and nutritional plans. Call now 614-555-1122. Text STOP 2 end.

This week's specials at BIG Video: Mission Impossible 3, Monday; Rocky 7, Thursday. FREE popcorn with rental for you by showing this text. Text STOP 2 end.

Dan James Auto Mall! Wednesday from 11 am to 3 pm buy a new Volvo or SAAB and we'll offer $1500 in savings! Safety award winning new model. Text STOP 2 end.

Texas Auto Mall VIP members ONLY! Buy any pre-owned car in stock during the month of July and we'll provide a 2 year bumper to bumper warranty! Text STOP 2 end.

Visit our service department to get 25% off your maintenance and be entered in our "New Car Giveaway Sweepstakes" Southwest Auto. Expires 11/17 Text STOP 2 end.

NEWS! EJ Wines is extending their 20% discount offer until 6 pm this Friday. See additional offers on all our wines at www.ejwines.co.uk Text STOP 2 end.

A massive 50% off all sports products at Campus Leisure this weekend. Buy at www.campusleisure.biz. Ends at 7 pm on Sunday, quote code X13. Text STOP 2 end.

Dai Fashions pop up sale starts now! 24 hours only Use code POP10 and you'll receive a 24% discount on all items. Buy at www.daifashions.com Text STOP 2 end.

Plan for Mothers' day? Thoughtful gifts available at affordable prices. Visit us at 1 High Street or online. www.bluejaygifts.co.uk Text STOP 2 end.

Weather is looking cold and wet. Cozy up at The Rock Inn from $125.00 for 2 nights. Roaring fire, local ales, and spa. Call 613-555-1122. Text STOP 2 end.

Beat the line! Show this text at Oxygen Nightclub before 10pm tonight and go through the members' entrance. Free single shot with this text. Text STOP 2 end.

Celebrate great British cream tea at Mary's Tea Room. Free scone and dollop of Devonshire cream with every pot of tea today. 1 High Street. Text STOP 2 end.

FREE entry to the Brockville Paintball Centre through August. Just pay for any paintballs used. Keep the kids happy this summer. www.ypc.co.uk Text STOP 2 end.

Looking for something different to do this Friday? Come KARAOKE at Bar One. Two for One Cocktails EVERY Tues – Fri. See you there. Text STOP 2 end.

Time for Swimsuit Season? Book a trainer at The Gym before Sunday Nov 13th and get free nutrition plan (reg. $299) Call now 613-555-1122. Text STOP 2 end.

NEWS. Government raising insurance 15% – call to find how to beat the July 1st deadline. 613-555-1122 more at www.govinsurance.com Text STOP 2 end.

Your Van Insurance Renewal is due. Lower your premiums. Call today at 613-555-1122 and quote offer VANTXT www.yourvaninsurancerenewal.co.uk Text STOP 2 end.

RECALL NOTICE –Are you affected? Check out the list at www.yourford.com. Service specials until 11/17 Call Your Ford now 613-555-1122. Text STOP to end.

JOB ALERT! Please call Your Hired now at 613-555-1122 We have a number of exciting new jobs that may match your skills. www.yourhired.net Text STOP 2 end.

8 Text Marketing Mistakes

1. Making Your Content Feel Generic

Text marketing at its finest feels like a one-on-one conversation between you and every one of your subscribers. To make your content not feel like it is a form letter:

- Insert names or don't include a greeting - never address the subscriber as "customer" or "sir/ma'am."
- Use your business and product names whenever possible. Don't give a coupon for a "burger." Give one for the "Monster Madness Burger Deal."
- Use local details, such as a special celebrating your city team or community festival by name.

2. Not Double Checking Your Message

Having errors within your text message is very unprofessional. When you only have 160 characters to work with, you really need to make sure that you check your message as you are writing it and have someone else check it before you send it out.

- While one of your staff is proofreading, ensure they can identify the call to action. If that person can't recognize it, your message needs to be reworked.

3. Getting Off Topic

While a text message may be similar to Twitter, they are definitely not the same. Save your opinions and irrelevant thoughts for the tweets and focus your text marketing on highly relevant and valuable marketing content. If your message strays off target, you may find yourself facing opt-out problems or having your message treated as spam.

- If a customer opts-in for coupons, they will expect to receive coupons rather than general informational messages.
- The value of your offers must be relevant to your subscribers.
- Test multiple message types until you find what your customers respond to best.
- Listen to customer feedback and unexpected opt-outs to tailor future messages.

4. Using Text Speak

While it might seem easier to fit more characters into your message by using short form words, don't do it. You won't look cooler to the Generation Y customers if you try to use text speak. It is more likely that many of your customers will not be able to crack your code and your message will look like spam and be deleted.
- Avoid hype, slang and abbreviations.
- Leave out anything that might seem too slick and promotional.
- If you wouldn't write it on any of your other promotional material, don't use it in text.
- Avoid symbols, dropped vowels, and abbreviations.

5. Not Having An Appropriate Schedule

While sending too little text messages will cause you to miss the chance to build a responsive list, going to the opposite extreme and sending multiple daily texts with no real value will alienate your subscribers and lose traction. In both cases, you are guilty of poor timing.
- You should create a schedule of one message every three to seven days - enough to stay top-of-mind, but not enough to be disruptive.
- If you don't have a relevant message, don't send one. Never send a message because of the schedule; it is better to miss a day than waste your customer's time.
- Include "opt-out" codes in all your messages. Some subscribers are more forgiving of over-frequent contact if they see you've given them an out.

6. Not Building For A Future Relationship

If you fail to let your customer know that you will be back in touch, either during the opt-in procedure, or in subsequent messages, you will leave your customer surprised and angry when you send them future messages.

- Respond to your initial contact with a second call to action, you can also move them to other platforms by collecting email addresses or other details.
- Pay attention to your subscriber list - especially your response rates.
- Include a short note about future promotions, and your opt-out code, in all your messages.

7. Sending Your Texts At The Wrong Time

Research has shown that cell phones are used most between noon and 6 p.m., but this does not mean that these are the right times to contact your audience. You must research and know the cell phone behavior patterns of your audience. To start with, test by sending your texts during the late morning through to mid-afternoon periods. But...

- Avoid dinner or generally 'private' times of day.
- Don't text during drive time. When customers are commuting to work, they're more likely to ignore a message than respond. Plus it's dangerous.
- Don't program your message to be delivered during times normally associated with sleeping.

- Remember that different industries may require different times of day. A night club may choose to send out their offers on a Saturday evening, while a coffee shop may choose to out their offers on Monday-Friday mornings, and Restaurants may want to send out messages at 3:30-4:00 when hungry workers are preparing to go home.

8. Not Testing Your Marketing or Segmenting Your Lists.

If your messages are overly aggressive or too frequent, you risk having your customers consider your messages to be spam and tune you out. You must do some research to see how many text messages you should send out by reviewing what other businesses in your industry and local market are doing. Review your messages with a select list of customers to see what wording generates the best response and consider creating multiples lists so you can test the response rates for messages worded in different ways. If you want to send out messages every day of the week, consider creating a different list for each day or by different marketing segment.

Since 2007, McDonalds in Germany has used text message marketing to build more than 10,000 subscribers and achieved response rates up to 29%.
Mobile Marketing Association

Sprite prints unique mobile codes under their bottle lids, and gets 750,000 text message entries... in one month alone.
Mobile Marketing Association

20 Successful Text Marketing Campaigns

So far, we have covered in-depth all of the theory of how to run a successful text marketing campaign. Success will now come from how you take the theory and put it into practice.

The purpose of this next session is to present a number of different success stories in order to help you brainstorm ideas on how you can use text message marketing to boost your profits.

You should refer back to these cases routinely. As your experience grows, your perspective will change. You will find new ways, discover new ideas, and grow your responsive customer database.

Campaign #1 – Back To School

Columbia College leverages traditional media and gets an immediate response.[1]

Columbia College in Jacksonville, Florida was looking to stand out from the other college radio spots and connect with their target market to fill some of their spring classes. The challenge was that there were only 3 weeks until classes began.

Their target clients were adults 24 years old and older that were looking to finish a degree or pursue a Master's degree. Columbia typically purchased radio spots to create a response, but because there were a number of similar ads from alternate universities already running, they decided to differentiate their offer by adding a text message call to action at the end of their ads.

The last 10 seconds of each radio ad simply asked the listeners to text the keyword DEGREE to a short code for more information. There was no further incentive. Each opt-in phone number was forwarded to the admissions department and the subscribers were personally contacted within 24 hours.

---

[1] Source – US Short Code Administration - http://www.usshortcodes.com/sms-marketing-case-studies/csc_case_ballyhoo_2.php

Results

The College set a goal of generating 10 qualified leads for this campaign. After the campaign was complete, they were pleased to find that they had generated 32 qualified opt-ins, and of these, 2 registered for spring classes and an additional 10-15 were estimated to be ready to start in the summer session.

Key Learnings

- You don't need to find a large prize, or deep incentive when you focus your campaign on a highly targeted, qualified prospect.
- When you forward a text message opt-ins immediately to a staff member that can respond, it provides greater opportunity to catch prospects at the instant they are in the right frame of mind to take action.
- You don't need to plan for months in order to get tangible results. A few weeks, or even a few days, can turn a flat campaign into a positive return.
- Text marketing can add to your traditional marketing by catching the additional prospects that are curious, but not quite ready to take action... increasing your overall response.

Campaign #2 – Text For A Cove.

Filmmakers Turn to SMS Marketing to Drive support... Protecting Dolphins From Slaughter.[2]

"The Cove" is an Academy Award winning documentary that follows activists as they infiltrate a heavily-guarded cove in Taiji, Japan. It is here that 20,000 dolphins and porpoises are slaughtered each year and sold as food – often labeled incorrectly as whale meat. The meat also contains toxic levels of mercury.

During the end credits, viewers were asked to text DOLPHIN to a short code. This would reach potential citizen activists and keep them informed about The Cove. After a subscriber texted the short code, they received a message with a link to sign a petition to shut down The Cove.

When "The Cove" won the Academy Award, the film makers held a sign on-stage asking viewers to text in. This was quickly cut off by the telecast producers, but in the days that followed quickly went viral on social media. Even Ellen Degeneres tweeted about it to her 4.3 million subscribers.

---

[2] Source – US Short Code Administration - http://www.usshortcodes.com/sms-marketing-case-studies/csc_case_cove.php

Results

During initial release, it was estimated that 9% of all movie viewers texted in to the campaign and 29% of subscribers signed the petition. "The Cove's" gained tens of thousands of supporters from theatrical release to DVD release, and it is still growing. The results are so positive that the media company is pursuing its own dedicated short code to support future projects.

Key Learnings

- Text Marketing is not only to send coupons and promotions. There are times when it is best used as a way for people to respond when they are not close to a computer. The results would not have been powerful if they had only placed a website address at the end credits.
- Text Marketing can have legs and contribute to long term project success
- The short character count of a mobile opt-in offer makes it easy to place in a social media message like a Facebook post or Tweet.

Campaign #3 – Catching The Fashion Bug

Women's Retailer Drives Opt-Ins with In-Store Incentive[3]

National women's retailer Fashion Bug wanted to grow their mobile consumer database, so they planned a unique text marketing campaign.

Through in-store signage, they prompted shoppers to ask a store associate for a secret text code to get an instant $5 coupon to use on their current purchase. Customers who texted in were also asked to opt-in for future text message offers.

Fashion Bug also used unique keyword and coupon codes to track the effectiveness of this campaign compared to other campaigns that they were running at the same time.

---

[3] Source – US Short Code Administration - http://www.usshortcodes.com/sms-marketing-case-studies/csc-case-study-fashion-bug.php

Results.

Within two weeks, Fashion Bug grew their database by 24%, or more than 50,000 mobile opt-ins. During the campaign, 3.5% of all consumers that came into one of their 700 stores joined the program, and opt-out rates continue to remain low. Fashion Bug finds that text marketing allows them to develop, test, and launch new initiatives that they can scale across their entire organization.

Key Learnings.

- Avoid using coupons or promotions that customers need to use on their next visit. Allow them to take advantage today.
- Consider using text marketing and your subscriber database to test new campaigns with the best opportunity for success before rolling them out to mass media
- The best source of new business is the group of people that you already have in your store. Catch their information while they are in front of you and not depend on your marketing driving action before they arrive, or worse… or pray for them to join after they have left.

Campaign #4 – Chuck E. Cheese

Family Restaurant Drives Email Opt-In Through Text Messaging[4]

Email marketing is a key element to the marketing success of the family entertainment and casual dining restaurant Chuck E. Cheese. Club members receive exclusive offers, news about activities, and other promotions regularly in their email in-boxes. In addition to the website opt-in process, the restaurant would use an in-store process that consisted of guests filling out a paper form. This process proved time consuming and produced many illegible email addresses... in addition to requiring staff taking the time to enter them into the system manually.

To solve this inefficiency, Chuck E. Cheese decided to automate the process by allowing the consumers to text their email address to a short code. Subscribers would then receive an immediate response with a welcome email and a coupon, rather than weeks later.

---

[4] Source – US Short Code Administration - http://www.usshortcodes.com/sms-marketing-case-studies/csc-case-study-chuckecheese.php

Results.

The new process now accounts for 5% of their entire daily sign up volume. More importantly, these new subscribers have an email open rate that is 10-20% higher and coupon clicks that are 8-10% higher than all other types of subscribers.

Key Learnings.

- Text messaging can be used to make inefficient processes easier and save time for both your customers and your staff.
- It is essential to market to your customers through multiple channels. You can use text messaging to build a mobile phone database as well as ask for their email addresses so you can market through that channel as well.

Campaign #5 – The 12 Days of Christmas

Route 66 Harley Davidson Uses Text To Harvest Business From Their Social Media.[5]

Route 66, a Harley Davidson dealership in Tulsa, Oklahoma, looked to use text message marketing to increase merchandise sales through the holiday season. They chose to target current customers and in-store customers to build their text marketing database. Customers were encouraged to text the keyword RT66 to a short code through the use of social media, in-store marketing and connecting to a database of previous customers.

Typically a slower time of year for sales, Route 66 offered a 20% discount on a different product from its merchandise each day for the 12 day campaign. This was promoted through a sequence of 12 daily text messages.

---

[5] Source - Mobile Marketer - http://www.mobilemarketer.com/cms/resources/case-studies/12164.html

Results.

The dealership recognized a significant increase in sales of the featured products during the campaign. It didn't matter whether it was lower-end or high-dollar items, increases ranged from 16% to 250%.

Key Learnings.

- The key to success for mobile coupon campaigns is urgency. If you make the discount significant, limit the time frame from which to use them.
- Use text message marketing to promote additional sales from current clients. Route 66 is primarily a Harley Davidson dealership, but they recognized that they could make additional sales to the same customer by presenting upsell items. What kind of back-end or upsell items can you market?
- If you are going to increase your marketing frequency, try to tie it to a bigger campaign/purpose like a holiday or perhaps a charitable cause.

Campaign #6 – Text 2 Win at the Baxter Avenue Morgue

A Haunted House uses Text To Scare Up Maximum Revenues.[6]

The Baxter Avenue Morgue, a Haunted House in Louisville, Kentucky, is only open for a short duration during the Halloween Season. Because of this, they had only 3 and a half weeks to bring in the most teenagers and young adults possible. They decided to run a text message call to action in commercial spots on target-audience friendly tv channels (ABC Family, MTV, VH1, Cartoon Network, AMC) and targeted Halloween programming.

Subscribers received a chance to win free tickets to the Haunted House each week, and the company pushed additional coupons and messages out to the database in the time leading up to Halloween.

The company also tracked the number of subscribers each advertising spot received so they could maximize their response by targeting the most responsive time slots and shows.

---

[6] Source – Mobile Marketer -
http://www.mobilemarketer.com/cms/resources/case-studies/11779.html

Results.

Thousands of subscribers texted in from the targeted areas of Louisville and Southern Indiana, and more than 50 percent opted-in to receive future messages from Baxter Avenue Morgue. This was particularly impressive to the company, as the value of the prize was relatively low ($40) and the commercial spots were only 15 seconds in length. They were also impressed with the ability to text out admissions specials on Halloween Evening, while their customers were out Trick or Treating and eager to attend.

Key Learnings.

- A contest doesn't need to have a high dollar value prize in order to be effective. It is more important to have a timely offer.
- When reviewing your campaigns, don't just stop at knowing the number of people that opted-in. Learn about what caused them to respond. Is there a particular time, or medium that created a better response? If there is, focus your advertising dollar where it will be best used.
- The benefit to text messaging is that it is mobile. Be cognizant that your customer may be ready and waiting on your doorstep for the right reason to take you up on your offer.

Campaign #7 – The Eagle Claw Takes Off

Bioline Uses Text Messaging to Sell Out at A Trade Show.[7]

Eagle Claw, a leading fishing equipment manufacturer decided to release its new product, Bioline, at the International Sportsmen's Expo. Bioline is the world's first biodegradable fishing tackle.

In an effort to be consistent with the environmentally responsible product, Eagle Claw decided to use Text Messaging as a "paperless coupon".

Expo attendees were encouraged to text BIOLINE to a short code in order to receive the coupon, which they could redeem at the exhibitor station. The strategy was to run the text message call to action for the entire 4 day event.

---

[7] Source – Mobile Marketer - http://www.mobilemarketer.com/cms/resources/case-studies/8930.html

Results.

Response to the text message call to action was so high that the product inventory was depleted by the second day of the event and the call to action had to be taken down. Executives analyzed all sales conducted during the expo and determined that 23% of all Eagle Claw product sales were directly related to the text marketing campaign, and that consumers were noticeably excited to engage in the product line through their cell phones.

Key Learnings.

- Use text messages to deliver key information and eliminate paperwork that consumers often bring home and put in the recycling. Not only do text messages end up staying on your customer's phone for a longer period of time, they can also forward it to their friends.
- Networking and Trade Shows are not dead... Consumers are just engaging you in a different format than they used to. Get more creative and use text as a tool to connect and make face time more productive.

Campaign #8 – Legends in Concert.

Billboards Help Concert Venue Attract Tourist Business. [8]

Las Vegas based production company, Legends in Concert, wanted to measure the effectiveness of their billboard campaigns, and build a text message subscriber base to increase attendance for their events in Myrtle Beach. With Myrtle Beach attracting 15 million annual tourists, they knew that the key to their success was connecting with this target group.

Legends in Concert used unique keywords on each of the 5 billboards purchased to measure location effectiveness and responded to opt-ins with a unique coupon code to measure conversion.

After opt-in, a subscriber would receive an automated staggered series of text messages based on the approximate length of their visit. Each message would build in urgency (i.e. Hurry, last chance offer) and provide discounts on show tickets.

---

[8] Source – Mobile Marketer - http://www.mobilemarketer.com/cms/resources/case-studies/8819.html

Results.

While building their text message subscriber database, Legends were also able to track the effectiveness of a media that they previously couldn't measure. The combined billboards had an average conversion rate of 16% and a 1200% return on investment. Interesting enough, their most expensive billboard had the lowest conversion rate.

Key Learnings.

- Don't just think of a campaign as a single message. You can automate and send a sequence of messages in order to build interest and lead your prospective client into a well-educated purchase.
- Consider the ability to split-test and track multiple factors in order to focus your campaign on the most profitable activities.
- Consider that the most expensive marketing choices could be your least effective.

Campaign #9 – Southern Shows Wins in Guerilla Fashion

Mobile Coupons and Contest Creates a Win For Women's Show.[9]

Southern Shows, a women's show in Charlotte, North Carolina, used text messaging to drive ticket sales and measure the return on investment for traditional advertising mediums.

Subscribers were given a chance to win a $250 shopping spree at the show for participating and every subscriber received a $2 off admission at the ticket counter.

Text to win calls to action were added to current radio spots, print ads and promotional materials… including tent cards. The show encouraged their exhibitors to place the tent cards in their stores, restaurants and offices. The biggest benefit of the tent cards was that it allowed more space to include detailed instructions on how to participate.

---

[9] Source – Mobile Marketer - http://www.mobilemarketer.com/cms/resources/case-studies/8801.html

Results

Based on the number of text participants, the coupon redemption rate for the show was 20% (much higher than any of their other media).

More surprising was that the tent cards outperformed the more traditional radio spots as the number 1 catalyst for action. The show was able to take this information and save on costs by allocating more resources to the tent card campaigns.

They also found that text messaging was much more effective than their email and social media campaigns as the messages did not get lost in the shuffle.

Key Learnings

- While simply advertising a keyword and short code can be enough for success in a campaign, you can achieve even greater results by including detailed instructions at point of participation.
- While text can replace your social media and email marketing efforts... it can also amplify your results and create more business by working together.

Campaign #10 – When's The Next Bus?

Orange County Transit's Automated Bus Schedules Save Thousands[10]

The Orange County Transit Authority in Orange, California faced a big problem with the costs of printed bus schedules and calls to the Customer Information Centre. Customer service calls cost the Authority close to $144,000 per month. They recognized that they needed a more convenient, cost-effective solution that got their customers the information they needed.

The Authority decided to use a dedicated short code in order to send schedule information directly to their riders. All a passenger needed to do was text the bus stop and route numbers to the short code and they would instantly receive a text with the pick-up times from that specific stop.

The Authority promoted this using signage in their buses and at bus stops, as well as creating a "Street Team" to visit stops and teach passengers how to use the service. They also created a video to show customers how to use the service.

---

[10] Source – Mobile Marketer - http://www.mobilemarketer.com/cms/resources/case-studies/8185.html

Results

While calls to the contact centre cost the Authority on average $2/call, the Text service came in at $0.10/call, a significant savings, and much better customer service. Call volume dropped as much as 40% and an estimated savings of $350,000.

They also found that their customers are much more tech-savvy than expected. Average monthly text volume is about 50,000 per month.

Key Learnings

- Sometimes a penny saved is a penny earned. Are there key ways that you can use text message marketing in order to improve customer service and more effectively allocate time for your staff to focus on generating more business?
- While your customer base is probably much more tech savvy than you give them credit for, it is always a benefit to have staff trained to be able to teach your customers how to use your text marketing service.

Campaign #11 – Text To Serve.

AnMarc Travel Uses Text2Chat For Superb Customer Care.[11]

AnMarc offers travel products and services to military families and travel management companies. They have more than 4.6 billion mobile users worldwide.

After having saved significant time and money moving their customers from service calls to online chat, AnMarc wanted to take the next step and eliminate the need for WiFi and laptops to talk to a travel agent. Many of their clients need to book travel plans at the last minute, and text allowed them to do so anytime and anywhere.

AnMarc's belief is that the travel industry is getting smaller and the needs of the traveler are getting larger. In providing access to a live agent via text chat, they would not only empower the traveler, but also decrease overhead involved with fixing a problem after the fact and avoid frustration and resentment built by the avoidable situation.

---

[11] Source – Mobile Marketer -
http://www.mobilemarketer.com/cms/resources/case-studies/8062.html

Results

Less than two months from initial deployment, more than 25 percent of AnMarc's live chats were converted from PC to text. With an 80 percent retention, this helped to save money on staffing costs, as mobile live chat lets customer service agents handle up to four customers at once.

In addition, AnMarc envisions this technology can provide in-flight customer service for the airlines, meaning that they could represent the airline by assisting their traveler in making changes to existing reservations while in flight.

Key Learnings

- With the 'instant access' mentality cell phones bring, businesses must continue to shift their ways to assist customers.
- Create customer service plans that mirror the modern, mobile consumer.
- AnMarc was very surprised to see the adoption rate of mobile chat versus PC-based Internet chat happen so quickly, and was also surprised that users tended to use mobile chat even though they were within reach of their PC.

Campaign #12 – Radio Talk Show Uses Text To Connect

Dawson McAllister Taps Text Marketing To Interact With His Core Audience.[12]

Dawson McAllister is a youth radio talk show host that was seeking to add a level of interactivity to the standard call-in radio experience. To this end, he has found text messaging to be crucial in reaching this goal.

A popular segment of his show is peer-to-peer advisor. During this segment, a caller's issue is thrown out to the audience, who can offer advice through the use of text, Twitter, Facebook, or web forum (text is most popular). They typically pull 8 to 10 of the text messages that come in to read on air.

The radio show treats mobile as an interaction tool. They understand that their audience doesn't trust easily. Texting can be a very non-confrontational way to help build that trust up over time.

Mr. McAllister also encourages teens with a specific issue to text. An immediate automated text response containing

---

[12] Source – Mobile Marketer - http://www.mobilemarketer.com/cms/resources/case-studies/7733.html

information and Web site links for that issue is sent to each inquiry.

Results

Radio listeners are typically on the go. Text message and social media from mobile is a very effective way for Dawson to deal with very personal issues in a private matter. He also keeps in contact with the listeners off air via regular text alerts, which listeners can sign up for on the Dawson McAllister Web site.

Mobile interaction on the show happens in a number of ways – via text in functionality, text alerts and automated text for info services.

Key Learnings

- Use text to connect with your audience during group presentations.
- You can provide real-time feedback in a group or presentation environment at the exact time the customer needs it.

Campaign #13 – Success Through Segmentation

Carrabba's Italian Grill Uses Text To Build Business In Multiple Key Areas[13].

Carrabba's wanted to test reaching customers through multiple points at different times of the day. Early dining (4-6), Late dining (8-10), Carside Carryout, Sunday lunch, and Happy Hour are some of the different campaigns that were used.

Additionally, Carrabba's wanted to further engage customers and learn more about their interests while deepening brand awareness and technological advances.

With a goal of a 5 percent redemption rate, each participating Carrabba's location had the ability to grow their own database and select specific business touch points where they wanted to grow their sales.

Calls to action varied among each campaign that was used. Different offers were used to resonate with the customer and encourage repeat visits and increased traffic to restaurants.

---

[13] Source – Mobile Marketer - http://www.mobilemarketer.com/cms/resources/case-studies/7233.html

Results

There were 443 participants, accounting for a 35 percent redemption rate. This launched the Carrabba's customer appreciation group and allowed them to increase customer traffic by growing specific customer segments.

Key Learnings

- The key to running multiple campaigns and growing segments of customers in your business is to differentiate your lists and connect with them at the right time with the right message. Never just build one list.
- What are the different touch points and key buying patterns of your customers? Can you use segmentation to build areas of weakness?

Campaign #14 – Breaking Updates Mastered Through Text

The Globe and Mail Newspaper Uses Text To Keep Readers Engaged During G20 Summit.[14]

When world leaders gathered in Toronto for the G20 summit, The Globe and Mail wanted to offer breaking news updates and analysis of the issues to their readers.

The Globe needed a campaign that would work nationally and would allow for interaction in a way that differentiated it from other publications. They offered mobile alerts to allow the reader to stay up-to-date with the latest news as it happens.

Through the course of the G8/G20 Summit and for several days leading up to it, subscribers received breaking news updates relating to the event. Each text message contained the news headline along with a web link to the article that navigated readers back to the Globe and Mail's website. At the end of the campaign readers were asked if they wished to stay opted in for future news alerts by text or if they wished to provide their email address to have news sent to their in-box.

---

[14] Source – Mobile Marketer -
http://www.mobilemarketer.com/cms/resources/case-studies/6800.html

Results

- 53,000-plus messages were sent as part of the campaign
- 52% of text subscribers gave their email address to receive future news subscription alerts
- 89% of alert subscribers signed up for the alerts online

Key Learnings

- Think of how you can use text messaging to differentiate yourself from the competition. The Globe and Mail had an opportunity with the summit to get the news to their subscribers faster and more efficient than the competition.
- Can you tie your marketing campaign to an urgent customer need or an event?

Campaign #15 – City Asks You To Relieve Yourself

Winnipeg Health Authority Capitalizes on Privacy.[15]

The Winnipeg Health Authority needed to reach youth ages 18-24 in order to raise concern and awareness about the incidents of the sexually transmitted diseases gonorrhea and chlamydia. The goal was to get people to get tested for sexually transmitted infections, increase awareness of incidents and remove false perceptions about testing.

The campaign media buy focused on public transit and also included Facebook ads and in-theater calls to action asking subscribers to text the keyword PEEINACUP to a short code or go to http://www.peeinacupwinnipeg.ca for clinic locations and a chance to win $1,000.

Through text messaging and the mobile web, the company created a channel for youth to gain instant and private access to important information about a sensitive topic.

---

[15] Source – Mobile Marketer - http://www.mobilemarketer.com/cms/resources/case-studies/6609.html

Results

Over the course of the four-week campaign, more than 10,000 people visited the Web site and 825 people entered the contest, which is 1.2 percent of the total target population in Winnipeg.

Key Learnings

- The biggest reason that the campaign worked so well was not the fact that there was $1,000 up for grabs, but rather the fact that the cell phone is a person's most intimate device.
- If you have sensitive material that may be embarrassing for a customer to see, deliver it via mobile. A customer is more likely to engage because they feel they can be anonymous and their information is secure.

Campaign #16 – The Green Loyalty Card

Farmer's Foods Uses Text Marketing to Drive Store Traffic.[16]

Farmer's Foods is a family-owned and operated grocery chain, with nine locations, seven in Virginia and two in North Carolina. They wanted to build a membership base using text messaging instead of a rewards card to drive in more store traffic to increase overall sales.

Farmer's redesigned their web site to carry the message, used print ads, banner ads, weekly flyers, and a limited amount of radio and TV. They also used a life-size cutout of its mascot in-store to attract more attention and carry the message.

After opting-in, the subscriber receives weekly discount coupons and is automatically entered into monthly contests for shopping sprees.

---

[16] Source – Mobile Marketer - http://www.mobilemarketer.com/cms/resources/case-studies/6024.html

Results

To date Farmer's Foods has acquired more than 1,900 members via mobile. Its coupon redemption rate ranges between 8 percent and 15 percent, depending on the product offered.

At first, Farmer's Foods noticed its opt-out rate was a little higher than it liked. So, the company added a monthly contest with prizes such as a chance to win a $100 shopping spree.

All active members were automatically entered. At the end of the month a winner was chosen and a message was broadcast to all members of who the winner was.

Key Learnings

- Farmer's Foods recently broadcast a promotion that drove members to the Web site for a printable coupon. Following that broadcast, several members unsubscribed immediately. They believed that the opt-out was because the fulfillment was not on a mobile platform—for the members to participate; they had to go to a computer to print out a coupon.
- This shows that in order stay on course when using mobile rewards, do not send the customers to another platform for redemption.

Campaign #17 – Hair Salon Cuts Their No-Shows

Headrush Salon Chooses Text Messaging To Keep Their Appointments In Their Chairs[17]

Headrush Hair Salon was looking to reduce the amount of 'no show' clients who may have forgotten the date or time of their appointment.

With a customer base of 4,700 clients in the East Sussex area, this was a major problem as it could leave a stylist in the salon with sometimes up to 2.5 hours of spare time. This is clearly bad for business and profits

They selected a text messaging service that allowed a way to communicate with clients and remind them of their upcoming salon appointments. All they needed to do was set the salon computer to send the messages on the day before each appointment. This could be easily done by staff randomly throughout the day or programmed well in advance to automatically deliver to the customer at the appropriate time.

---

[17] Source – Mobile Marketer - http://www.mobilemarketer.com/cms/resources/case-studies/5744.html

Results

Headrush calculates that its number of no-shows has been reduced by approximately 70 percent and expects this number to improve further as they obtain more cellular subscribers from their clients list.

Key Learnings

- Appointment cards can be lost and reminder phone calls can be intrusive and take valuable man hours. Can you use text messages as an automated system to remind a customer of service calls, appointments, meetings, etc.

Campaign #18 – Coldwell Banker's Mobile Showroom

Real Estate Broker Uses Text To Streamline Processes[18]

Hoping to streamline the process of buying a home, and give individuals selling their home more choices, Caldwell Banker placed text calls to action on sign riders. Each rider had a unique house code that individuals can text in to get specific information about the house, including a link to a mobile web site, and a way to text in for an agent to contact them.

This brokerage was focused on the highest level of customer service, and wanted to ensure that sales associates had superior services and technology to meet the needs of clients. They found that text marketing technology gave buyers immediate and comprehensive information about listings in a very convenient way, and gave their agents a truly competitive edge in attracting potential buyers.

---

[18] Source – Mobile Marketer - http://www.mobilemarketer.com/cms/resources/case-studies/5504.html

Results

They found that individuals want to use mobile to browse homes and be more in control of the home buying process, but still text in AGENT to have an agent call them. Text assisted in the process, but does not tremendously cut into the role of the real estate agent.

Key Learnings

- Text messaging can assist in higher end product sales. Use text messaging to deliver specific information and shorten your time required in front of a customer.
- Remember that some phones cannot handle mobile web pages, so ensure that you have an alternative that will allow a customer to get in personal contact with a representative on all campaigns.
- Consider using text messages to create a "mobile business card".

Campaign #19 – Pizza Hut Encourages Friend Forwarding

Restaurant Stands Out With Contest That Has Legs[19]

In an effort to make its pizza and pasta dishes stand out among other brands in the Pittsburgh area, Pizza Hut used a text messaging contest. This contest featured prizes such as the first prize of free pizza once a month for a year and a second prize of chicken alfredo. Other prizes included Pepsi or a large pizza for forwarding to friends.

When a subscriber forwarded Pizza Hut's text to five friends, they received a free two-liter of Pepsi and by forwarding to 10 friends they received a free large pizza. This further drove participation and put individuals in control of their ability to win prizes.

As well as friend forwarding, Pizza Hut used television spots promoting a short code and keyword call to action.

---

[19] Source – Mobile Marketer - http://www.mobilemarketer.com/cms/resources/case-studies/5435.html

Results

Pizza Hut saw more than 12,000 entrants in the first two weeks. More than 3,000 texted in during the commercial and the rest of the entrants were driven by the friend-forwarder.

The combination of a great prize combined with other compelling consolation prizes that are easy to attain and friend forwarding led to much higher than anticipated participation numbers.

Key Learnings

- Tying the right incentive or collateral to a campaign or contest will encourage your customers to pass the information on to their friends and family.
- While it is always good to encourage referrals, please ensure to use this tactfully. If your approach using this medium becomes too aggressive, you may find yourself experiencing a larger rate of opt-outs from your subscribers.

Campaign #20 – Zorbaz The Great

Restaurants Consumer Base Drives Change From Local Carrier[20]

Like previous campaigns, Zorbaz turned to text messaging to build their guest loyalty program... The Zip Club.

Zorbaz used email to contact their current email ZIP members. The email explained the new texting service and offered ZIP members a link directly to an online landing page, to opt-in and specify their neighborhood location.

The goal was to segment their loyalty subscribers into their location of choice, and build location traffic based on specials focused on that area's demographic.

Text messages almost always promoted a discount or special on food or beverage, or inform or remind guests about a special event, such as the Ugly Zweater Party, Winterfezt, Oyzterfezt, karaoke, live music and other events.

Each location's manager had control over when the message was sent and the content of the message.

---

[20] Source – Mobile Marketer - http://www.mobilemarketer.com/cms/resources/case-studies/5138.html

Results

The Zorbaz text messaging program grew rapidly from its inception.

But more surprising...

Sometimes short codes are not carried by all carriers. Many Zorbaz guests were frustrated when they learned they could not be a part of the text program because their carrier did not offer short code text as a service.

The demand from the carrier's customers to allow this service became so great that the carrier's representatives called the text service provider to find out what was causing this stir. The provider explained the Zorbaz text program and how popular it had become About three months after the issue was brought to light, the carrier officially began to offer short code text messaging for the Zorbaz short code.

Key Learnings

- Zorbaz found that text messaging helped maintain top of mind awareness and increased their purchase frequency.
- Send out intelligent and well-thought messages to a highly targeted customer and make sure the message includes something desirable.

Text Marketing 102

The Technology

When establishing a text message marketing program, you are going to be faced with some choices in regards to the technology and software available. In order to make an informed decision, you are going to need to know the basics behind the technology.

The technology behind text messaging is SMS (short message service). Cellular phones are constantly sending and receiving information. Even when the phone isn't in use, signals are sent and received from a cell phone tower or control channel. The control channel maps the path for text messages. When a message is sent, it first must go through the nearby tower and then the SMS center. The SMS center receives the message and sends it to the appropriate tower closest to the location of the cell phone and then to the destination. The short message service formats the message in a way that it is able to travel through this process and be interpreted by the cell phone.

In text message marketing, there are two methods used to make this technology work, SMPP and SMTP.

Transfer Method 1 – SMTP Protocol

SMTP, or "Simple Mail Transfer Protocol", can be commonly referred to as "Email To SMS". SMTP was first published in 1982 as the primary means to send email messages. In a bid to keep subscribers, cell phone carriers decided to give every customer their own email address using their phone number and a carrier web address.

Some advertisers caught on to this as a cheap way of sending someone an "SMS" message. To send a text message via the SMTP method, all you need to do is attach their phone number to the beginning of the carriers SMTP email address and send it like any other email.

Those advertisers found that this was a great way to start because the only real cost are the internet charges and the time it takes to figure out what each carriers domain were (i.e. @mobile.mycingular.com). Sounds good? Why wouldn't you want a cheaper and easier way to do something? There are a few reasons why cheaper is definitely not easier when it comes to text marketing.

Transfer Method 2 – SMPP Protocol

Technically called "Short Message Peer-to-Peer Protocol", SMPP is the authentic and preferred method of text message marketing. This method routes messages through the carrier's cell network and is the same method as you use when you are sending text messages back and forth between friends.

This is the preferred method as providers and carriers build relationships so the carriers understand that the messages sent are following best practices and assist in the delivery. To get started you will probably incur a connection, set-up and a per message fee to send each text message.

5 Reasons You Want To Choose SMPP Over SMTP

While SMTP is the cheaper way to market via text messaging, there are very distinct advantages to using a SMPP vendor. These are:

1.   **Deliverability** - SMTP is not a direct connection to carriers. The carriers do their best to block commercial messages through SMTP as a majority of their complaints come from this method and in return, messages are much more likely to not be delivered. Even worse is that these messages may not be delivered and you will not even know about it. Carriers and suppliers cannot produce reports for SMTP marketing. With SMPP, you can receive reports on successful and failed deliveries, and why a text message failed. Also, unlike U.S. carriers, there are few international carriers that offer an email address domain.

2.   **Timeliness** - SMPP offers businesses priority routing and the ability to deliver your messages exactly when you want them delivered. SMTP messages are extremely slow and may take 24-48 hours to deliver. SMTP system is not a bulk delivery system. You can't text all of your customers at once and some systems have you enter each number manually. Some programs deliver the messages in sequence one after the other until done. This can be a huge issue depending on what the message is. Timely delivery for text marketing is critical. If you are sending out a lunch special and it isn't delivered until 9 pm it doesn't do you any good.

3.   **Reliability** – In order to prevent spam, carriers change their email domains. What happens to your campaign when this happens? All of a sudden your messages stop getting delivered. When you send messages via SMPP, your connection will always be the same and you won't have to play this guessing game.

4.   **Lack of Communication** - When you send a text via SMTP, your customer does not have the ability to respond to it. What happens when you want a response? You will not have the ability to poll your database or give them a chance to win something by responding. SMPP offers full

two-way text messaging so you can create a dialogue with your customers and get important feedback. Also, there is no ability to get a short code for SMTP. Nobody can text into your brand to interact with you. The only thing you can do is send a one-way message out.

5.	**Legal** – SMPP is the legal way to send text marketing messages. SMTP does not give you the ability to have a service provider track opt-ins. You open your business up to liability. The SMTP gateway was only built for personal use, and carriers have made it very clear that any use of this gateway for commercial purposes are in direct violations of their terms of service. There is also no way for someone to quickly and easily remove themselves from your mailings or reply back with an unsubscribe request. SMPP protocols are regulated by the wireless carriers, all the necessary regulation are in place. SMTP is not regulated and is a loophole in the carrier regulations. If you choose SMTP, ensure that you have the systems in place to track opt-in, opt-out, and notice and privacy notifications manually. This is something that is done for you by a SMPP provider.

When comparing SMPP versus SMTP and selecting a text message provider, it would be in your best interest to choose a provider that offers the true SMPP protocol as their delivery method. It will save headaches in the long run if you don't have to worry about all the disadvantages of SMTP.

What You Need To Know About Short Codes

What Is A Short Code?

The most common and preferred method to building your subscriber lists is to get users to text the advertised keyword into your short code. Like "Text COUPON to 54321 to receive our special of the week".

A short code is simply a shortened 4 – 6 digit telephone number that is used by businesses for sending text messages. A business may purchase their own short code or share a short code with other businesses. If a business purchases their own short code it is considered a dedicated short code, in that it is dedicated solely for the purpose of that business to be able to text message their contacts. Short codes are leased through the CSCA (Common Short Code Administration) and then provisioned through the wireless carriers and will generally require the assistance of an aggregator or application provider.

The 3 Types Of Short Codes.

Shared

Shared Short Codes are typically owned by a text message marketing provider. The provider will then form text message packages that offer their customers the ability to use their short code, have their own keyword (or keywords), and an allotted amount of text messages a month for a specified fee. Shared short codes are an excellent way for any size of business to save money on text message marketing.

When sharing a short code the keyword is what would differentiate business lists. For example:

- Tata's Pizza and Avalon Spa share the short code 555444
- Tata's uses the keyword "PIZZA"
- Avalon chooses to use the keyword "SPA"
- Both businesses send text messages to their customers cell phones
- All clients get a text message from the short code 555444
- When customers reply they must type the keyword a space and then their message.

- The customer message is then sent to the appropriate business based on whether they replied "PIZZA" or "SPA"
- "PIZZA" responses are delivered to Tata's, and "SPA" messages are delivered to Avalon.

Vanity

Vanity Short Code are Dedicated Short Codes that spell out a word (for example OBAMA – 62262).

Dedicated

A dedicated short code is your own common short code, which you lease from the Common Short Code Administration (CSCA) and setup through an aggregator or service provider. With this, your keyword availability is unlimited. Dedicated Short Codes are ideal for organizations needing a large number of keywords and organizations that do not want to share a Short Code with other businesses.

Pros and Cons of Shared Short Codes

PROS

1. **You can start quickly**: Sharing a mobile marketing company's short code is the fastest way to get started. Typical campaigns can be live within a couple of days.
2. **Low Investment**: Using a shared short code is much, much cheaper than leasing your own short code and you will generally just pay a monthly fee for the keywords and the number of messages that you use. Some typical costs include:
   - 1,000 messages per month averages about $50 per month
   - Extra messages added average between $.04-$.07 per message
   - Most providers include 1 keyword, extra keywords run between $8-$25 per month
   - Setup fees vary depending on the scope of the project.

CONS

1. **Fair user experience**: Your user experience may be a little more complicated because you are sharing your short code. On a shared short code you need to use keywords to associate responses with all of your programs. If you are asking someone to vote using A, B, and C and your keyword is FOOTBALL, you would need to get your consumers to reply

"FOOTBALL A," "FOOTBALL B," or "FOOTBALL C" so the system knows which campaign the voter is responding to.

2. **Brand confusion**: There is a very good chance someone else in your industry is using the same code you are, possibly even competitors. You have no control as to what companies have access to that code. If another business heavily advertises this code, your business may be associated with the other company's promotions. When the subscriber receives a text message, it comes from the five- or six-digit short code. If that subscriber is signed up to two different companies using the same code, he or she won't immediately know that it was you who sent the message. There is always the possibility that they want your texts, but not those from the other company. They could get upset and reply "STOP," and not realize they just removed themselves from your database.

3. **Keyword Availability**: Since the code is shared by multiple organizations, keywords can only be used by one organization on that particular short code. Because of this, some of the more popular keywords, like SALE, ALERTS, or DEALS, may not be available for your campaigns. To get around this you have to put your company name in front of it, increasing the number of characters your customers need to text (i.e. instead of just "JOIN" you'll need to use "JOINAVALON").

4. **Service Provider Risks:** There is always the risk that the short code may be shut down by a carrier because of someone else's abuse of the code. This may hinder your ability to run future mobile campaigns. Your service provider should take care

of this, but it is always prudent to secure a backup plan.

Outcome: Shared short codes are wonderful for small businesses. You businesses can get to market quickly, with little financial investment.

Pros and Cons of Dedicated/Vanity Short Codes

PROS

1. **Brand-friendly**: A company that has its own short code has the benefit of having a number their subscribers recognize right away who is sending them a text message, like 73775 (Pepsi). You can also avoid problems like in 2007. Senator Barack Obama decided to secure his own dedicated CSC (62262 which spells Obama) and Senator Hillary Rodham Clinton initially chose a shared short code (77007). Aides found that some user texted the wrong keyword to Senator Clinton's short code and the reply was information about a different business.
2. **Portable**: If you lease your own code, you can switch aggregators or service providers and not worry about having to reintroduce your subscriber base to a whole new code. There is no brand confusion that can typically happen.

CONS

1. **Having Your Own Short Code is Costly**: Having your own dedicated or vanity short code is a large investment. As well as the cost of the short code, most mobile service providers charge a monthly fee to host and maintain the code on a monthly basis and to provide the software platform to run your campaigns. There are also additional expenses to have someone help you apply for the code and see it through testing all the way to certification. When leasing a short code you will have to commit to a lease period: 3, 6, or 12 month terms are typical. Small businesses are not typically in a position to allocate the financial, technical, or personnel resources to manage their own dedicated short code. Here are some common fees to consider.

   - A random short code costs about $500.00 a month
   - A vanity short code is about $1,000.00 a month
   - A one-time set up fee of about $3,000.00
   - About $500.00 a month for platform use
   - A minimum of about $2,000 a month in message costs
   - A minimum 6 month commitment
   - Any charges from cell phone companies billed to the short code
   - Many may also charge a deposit of about $2,000.00

2. **Time to set up**: If you choose to pursue a dedicated short code, it will take a minimum of 6-12 weeks before you are able to meet all the

requirements and obtain all of the approvals necessary to start using your new short code. There is a complex approval process you must complete with the short code administration and carriers. If one or more wireless carriers turn down your application you will still be charged for your short code.

Outcome: If you decide that you require a dedicated short code, you should be prepared to pay and expect to jump through hoops and go through a lot of work to get started. You may want to consider using a shared short code while you are getting your dedicated one approved. This will allow you to get building your customer database right away and test which campaigns are effective before your dedicated code ready to use.

As you can see, there are a number of huge differences when comparing using a dedicated vs. shared short code. Shared short codes are the most common, most cost effective, and by far the easiest to obtain. When considering whether to choose a shared or dedicated short code, do your research and make sure you are aware of all the pros and cons. Chances are the comparisons will have you researching SMS shared short code providers.

Text Best Practices.

A Primer On Legal.

While this book was written with best practices in mind, any method of marketing has its own legal minefield. Text marketing is no different. From patent infringement based on technology used, to consumer protection, you always need to protect your business and resources.

I am not a lawyer, nor would I suggest that the research that I have done on this matter is complete. It would be my recommendation that you consult a qualified legal representative if you have any questions or concerns.

It would also be prudent to check out the following recognized authorities for their best practices and guidelines.

Federal Communications Commission (FCC) - The U.S. Federal Communications Commission regulates communications by radio, television, wire, satellite and cable. They protect the integrity of all mobile communications, including text messaging. In Canada, you should consult the CRTC (Canadian Radio and Telecommunications Commission). You should also research the authority in your area. The US FCC is a recognized authority and has some great basic guidelines to follow.

Federal Trade Commission (FTC) - The Federal Trade Commission is the USA's consumer protection agency. It collects complaints about companies, business practices, identity theft, and episodes of violence in the media. Its main function is to ensure text marketers adhere to proper business rules and conduct.

Mobile Marketing Association – The Mobile Marketing Association is the leading global non-profit trade association established to foster growth of all areas of mobile marketing. They establish mobile media guidelines and best practices. They have numerous case studies and research reports available for free. The first report you should read is http://www.mmaglobal.com/uploads/Consumer-Best-Practices.pdf.

CTIA - The Wireless Association – The industry trade group that represents a wide variety of interests on behalf of the international wireless telecommunications industry. Its members include all aspects of the mobile industry including wireless carriers and their suppliers, as well as providers and manufacturers of wireless data services and products. You can find valuable resources at www.ctia.org.

What Are The Risks If I Do Something Wrong?

The CTIA and Wireless carriers review text messaging marketing programs on a regular basis. If your program is found to break any of the rules and guidelines, you will be contacted by your vendor's mobile aggregator. If you are contacted about an infraction, you need to work closely with your vendor to complete the steps that need to be fixed. If you are working in an ethical manner and attempting to follow best practices, than typical infractions are usually related to improper language in your program's promotional materials.

If your violation is considered extreme, excessive or frequently recommitted, your text messaging provider will suspend your account (to protect both you and the vendor). The most extreme consequence is that you may have a class action lawsuit filed against you for violating your country's (or jurisdiction of authority's) consumer protection acts.

Most of these problems stem from consumer notices and complaints citing unwanted texts and violation of privacy. As a guideline, here are the most essential practices.

1. **<u>Text Message Marketing is always permission based.</u>**
Customers join your text marketing list by texting in a keyword to a number, a web form, or other method. This allows you to capture their phone number and by doing so they give you permission to send them messages. Sending

unsolicited messages will land your business in hot water and can cause you to lose your ability to send messages.

2. **In some areas, the Opt-out Procedure is more important than Opt-In.** Each outgoing message must contain instructions on how subscribers can opt-out. Examples include "Text 'STOP' to end" and "Thank you for joining our mobile coupon club. You will receive 1 message/week. Msg&data rates may apply. Text STOP to end." Your text marketing provider must be able to this automatically so the opt-outs do not need to be removed by hand.

3. **Provide Notice:** You need to provide consumers with the information they need to make choices about a marketing program.
- You must provide information about the products/services offered and the terms and conditions of the program
- You need to provide a HELP feature on your text messaging, or phone number for customer service.
- If you have a web page for users to join your mobile marketing list, the page should specify what types of communications the user will get, how often, and other important details. This should also be done at point of opt-in

4. **Customise Each Message For Your List and Practice Constraint:** Make sure you provide relevant and responsibly tailored messages targeted specifically to each marketing list. You should use information collected about users to customize marketing to their needs and interests, and handle any user information that's collected sensitively and responsibly. Failure to do this and your

users will consider your messages to be spam. Provide value and avoid sending irrelevant messages, avoid sending too many messages, sending messages at odd hours, or over aggressive tactics and harassment.

5. **Vet your text message vendor/provider**. Evaluate what their best practices are. What their customer service is like. What steps do they take to ensure data security? How are you protected from them spamming your customers? Detail your expectations and message delivery policies. Remember that no system is perfect. What security, procedures and policies do they have in effect that protects consumers' information? You need to ensure they have in place "reasonable technical, administrative, and physical procedures" to protect all user information collected.

Beyond this, you should keep in mind the following:

- All programs must comply with federal and state laws, as applicable
- "Msg & Data Rates may apply" must be clearly added on promotion/advertising
- Cost of premium or other fees must be clearly stated as well
- Advertising must note that participation requires users to be 18 years or older or have parental/guardian permission
- Selling mobile opt-in lists is prohibited
- The STOP and HELP keywords (not case-sensitive) should be featured in messaging to offer subscribers the opportunity to cancel or get more information about the program at any time.

- Document and save program opt-ins if you receive them in any method other than electronic, which should be saved by the text provider.
- Clearly communicate what people are signing up for (content, frequency, etc.) up front to ensure customer satisfaction
- Add your text marketing privacy policy to your website for easy access by program participants
- Work with your legal team to ensure your program offerings are legally compliant
- Follow through on your promise and send what you said you would send
- Provide a clear and simple method to request terms and conditions.

If You Stop Sending Messages

You should consider that an Opt-in text message subscription is not automatically a permanent list for you to market to. You should consider permission expired if you fail to send a message after a certain period. Please use these guidelines as best practice.

- If the only charge for the message is the standard rate charged by their mobile providers, consider the opt-in expired after eighteen (18) months.
- If you charge a fee to join the campaign, or customers pay for your information, consider opt-ins expired after six (6) months.

Patent Violations.

Patent lawsuits in the mobile industry happen all the time. The major brands are constantly tied up in court against each other in an attempt to gain the upper hand. Normally this has no impact on the small business owner, but certain patent holders are starting to pursue lawsuits with smaller service providers and small enterprises.

For example, if you send out a text message with a Web link in it to a mobile device, you are supposedly violating a "technology" that is a patent owned by Richard Helferich. He has several patents in and around that method of message delivery.

Ensure that you check with your text message provider to ensure that your business and marketing program is provided protection against possible patent infringements.

If You Need To Switch Software Partners – How to Migrate Your Mobile Subscribers

It is my recommendation that you always have one or two backup plans to ensure that your programs stay on track if any technology provider becomes unavailable. If for any reason you need to change companies, make sure you take care in how you do this so that it does not appear that you are in violation of SPAM regulations.

In this case, make sure you have records to certify that all numbers on the list were properly opted-in under the proper procedures and follow the migrating procedure outlined below.

- Using your previous provider's services, send one final message to your existing customers, letting them know that your future messages will come from a new short code. A message such as "Weekly coupons and alerts from Store XYZ are moving to a new number, 12345. Future text alerts will come from that number. Reply HELP for help, STOP to cancel. Msg&data rates may apply."
- Give your customers 48 hours to opt-out of the database if they wish to do so.
- After 48 hours, those who stayed in the database can be delivered to your new provider, who will import your database new platform.
- Once imported and opted-in, send a message to the clients introducing the new short code, such as "Welcome to 12345. Weekly alerts from store XYZ will now come from this number. Reply HELP for help or STOP to cancel. Msg&data rates may apply.

Glossary

Below is a short list of terms you need to know to be an effective Text Message Marketer. To view a more complete list of terminology, please visit the Mobile Marketing Association at www.mmaglobal.com

Acquisition Rate – Percentage of respondents who opted in to participate in a text initiative/campaign. Acquisition Rate is calculated by dividing the total number of participants by the total audience.

Aggregator – An organization or service that acts as a middleman between businesses, carriers, operators. Aggregators typically provide campaign oversight, administration and billing services.

Call to Action (CTA) – A statement or instruction that explains to a consumer how to respond to a message, promotion, or initiative. This is typically followed by a Notice

Carrier – A company that provides wireless telecommunication services. (i.e. AT&T, Bell, Verizon, etc.)

Click – The act when a subscriber interacts with an advertisement or actionable link.

Click-through Rate – The process of measuring success of an online advertising campaign made by dividing the number of users who clicked on an ad divided by the number of times the ad was delivered.

Common Short Code (CSC) – A short number sequence (typically 4 to 6 numbers) to which text messages can be sent. Customer's text messages to common short codes to access a wide variety of mobile content.

Common Short Code Administration (CSCA) – An organization that administers the common short code registry for a particular country/region. In certain regions local mobile carriers and aggregators are the administrators.

Dedicated Short Code – The process of running only one service on a common short code at any given time.

Double Opt-In – The process of confirming a subscribers wish to participate by requesting the subscriber to opt-in twice.

Free To End User (FTEU) – An FTEU program is any text marketing program to which the consumer receives text messages for which they do not incur any premium or standard charges from their mobile carrier.

Handset – Term used in reference to a mobile phone, device, or terminal.

Impression – each exposure of an ad to a consumer. For example, a text message, website view, video clip, etc.

Keyword – A word or name used to distinguish a campaign or targeted message with your text marketing service.

MMS Message – A message sent via a Multimedia Messaging Service that contains multimedia objects.

Mobile Originated Message (MOM) – A text message that originates from a mobile device.

Mobile Subscriber – a customer that subscribes to a mobile service that can send/receive messages.

Mobile Terminated Message (MTM) – A text message received on a mobile device.

Notice – An easy-to-understand written description of the information and data collection and privacy policies that may or may not be required of the company collecting and using information from a mobile subscriber.

Optimization – The process of modifying and refining an advertising campaign so that it will perform more favorably for the advertiser.

Opt-In – The process where a Subscriber provides explicit consent to receive information from a Marketer

Opt-out – The process where a Subscriber revokes consent. For example, replying to a text message with the phrase "stop".

Predictive Text – Intelligent software that makes typing words on a phone easier by using a built in dictionary to try to predict the words that the subscriber intends to compose based on the initial letters used.

Pull Messaging – Any content sent to the wireless subscriber upon request on a one time basis

Push Messaging – Any content sent on behalf of the marketers to a subscriber at a time other than when the subscriber requests it.

Royalties – A fee paid by a content aggregator/ service provider/ carrier to the content owner for use or repurpose of content or licensed processes.

SMS - SMS stands for Short Messaging Service, or text messaging. Pretty much every phone has it and it allows the delivery of short messages (160 characters) between mobile phones and other handheld devices.

SMS Marketing Service - providers of automated messaging services. SMS marketing services provide the technology platform and all features necessary to conduct text marketing campaigns at a low monthly rate.

SMTP Messaging – This stands for "Simple Mail Transfer Protocol" and is basically a method of sending text messages using email technology. Beware of companies that use this method for their means for getting text messages out. Text messages sent by this method are subject to email spam filters and run the risk of not landing on subscribers phones.

SMPP Messaging – This stand for "Short Message Peer-to-Peer Protocol" and is the preferred method for sending text messages. They use "short codes" or short numbers and send your texts directly through the cell phone carriers such as Verizon or AT&T to get your messages out. A text sent through this method is treated just like a text from one cell phone to another.

Targeted SMS Marketing - using various criteria to target a sms message to a specific demographic, location, or group. Businesses use targeted text marketing to increase their ROI.

TCPA - This acronym stands for the Telephone Consumer Protection Act. This act was signed into law in 1991 and protects consumers from receiving annoying telemarketing messages, including text message spam. Watch out, we're talking law, the kind of law that sends you to jail if you violate it.

Vanity Short Code - a type of dedicated short code that a company specifically requests. A vanity short code typically spells out a company's brand, name, or an easy to remember word associated with the business.

Wireless Spam (Unsolicited Messages) – Push messaging that is sent without a confirmed opt-in (big no-no)

You don't have to do it all alone.

There is no other business in the world that is exactly like yours. With this in mind, there is no perfect strategy that is guaranteed to work for your business. You must commit to the ongoing process of testing and adjusting to find out the most effective formula for you.

Let us help you shorten the learning curve. We can help. What strategies should you use? How do you measure success? How do you incorporate your strategies into your business process and manage effectively?

We can help guide you through these questions,
In business, time really is money. The longer it takes you to adjust to the new reality, the greater the opportunity for missed sales.

Visit us today at www.townshendconsulting.com or call us at 1-613-340-3247 to find out more about how we can help.

To your success,

Peter Townshend

About The Author...

Consultant, Speaker, Strategist, and Sales Expert, Peter Townshend has spent his entire career in the trenches working with entrepreneurs. Throughout his career, Peter's specialty has been walking into challenging situations and creating rapid positive change. Frustrated with how the new technologies have caused businesses to be slow to adapt, Peter has committed to becoming the catalyst for his clients growth in Mobile Marketing and a guide for change.

You can reach Peter directly at peteretownshend@sympatico.ca or via text message at 613-340-3247.

www.ingramcontent.com/pod-product-compliance
Lightning Source LLC
Chambersburg PA
CBHW071156200326
41519CB00018B/5244